Science in History

J. D. Bernal

In 4 volumes

Science in History

Volume 2: The Scientific and Industrial Revolutions

106 illustrations

The M.I.T. Press
Cambridge, Massachusetts

First published by C. A. Watts & Co. Ltd, 1954
Third edition 1965
Illustrated edition published simultaneously
by C. A. Watts & Co. Ltd and in Pelican Books 1969
Copyright © J. D. Bernal, 1965, 1969

Designed by Gerald Cinamon

First MIT Press paperback edition, March 1971
Eighth printing, 1986

ISBN 0 262 02074 2 (hardcover)
ISBN 0 262 52021 4 (paperback)

Library of Congress catalog card number: 78-136489

Contents

Acknowledgements

This book would have been impossible to write without the help of many of my friends and of my colleagues on the staff of Birkbeck College, who have advised me and directed my attention to sources of information.

In particular I would like to thank Dr E. H. S. Burhop, Mr Emile Burns, Professor V. G. Childe, Mr Maurice Cornforth, Mr Cedric Dover, Mr R. Palme Dutt, Dr W. Ehrenberg, Professor B. Farrington, Mr J. L. Fyfe, Mr Christopher Hill, Dr S. Lilley, Mr J. R. Morris, Dr J. Needham, Dr D. R. Newth, Dr M. Ruhemann, Professor G. Thomson and Dona Torr. They have seen and commented on various chapters of the book in its earlier stages, and I have attempted to rewrite them in line with their criticisms. None, however, have seen the final form of the work and they are in no sense responsible for the statements and views I express in it.

I would like also especially to thank my secretary, Miss A. Rimel, and her assistants, Mrs J. Fergusson and Miss R. Clayton, for their help in the technical preparation of the book – a considerable task, as it was almost completely rewritten some six times – and its index.

My thanks are also due to the librarians and their staffs at the Royal Society, The Royal College of Physicians, The University of London, Birkbeck College, The School of Oriental and African Studies, and the Director and Staff of the Science Museum, London.

Finally, I would like to record my gratitude to my assistant, Mr Francis Aprahamian, who has been indefatigable in searching for and collecting the books, quotations and other material for the work and in correcting manuscripts and proofs. Without his help I could never have attempted a book on this scale.

J.D.B.
1954

Acknowledgements to the Illustrated Edition

For the preparation of this special illustrated edition of *Science in History*, I must thank, first of all, Colin Ronan, who chose the illustrations and wrote the captions.

I should also like to thank Anne Murray, who has been responsible for correlating all the modifications involved in producing a four-volume version and for correcting the proofs.

Finally, I thank my personal assistant, Francis Aprahamian, who advised the publishers at all stages of the production of this edition.

J.D.B.
1968

Note

In the first edition of this book, I avoided the use of footnotes. A few notes have been added to subsequent editions and are marked with an asterisk (*) or a dagger (†) (if there is more than one footnote on a page). The notes have been collected together at the end of each volume and are referred to by their page numbers.

The reference numbers in the text relate to the bibliography, which is also to be found at the end of each volume. The bibliography has eight parts that correspond to the eight parts of the book. Volume 1 contains Parts 1–3; Volume 2 contains Parts 4 and 5; Volume 3 contains Part 6; Volume 4 contains Parts 7 and 8.

Part 1 of the bibliography is divided into three sections. The first contains books that cover the whole work, including general histories of science. The second section contains histories of particular sciences and the books relevant to Part 1. The third section lists periodicals to which reference has been made throughout the book.

Parts 2, 3, 4, and 5 of the bibliography are each divided into two sections. The first section in each case contains the more important books relevant to the part, and the second the remainder of the books.

In Part 6 of the bibliography, the first section contains books covering the introduction and Chapter 10, the physical sciences; and the second section, Chapter 11, the biological sciences.

Part 7 of the bibliography contains books covering the introduction and Chapters 12 and 13, the social sciences.

Part 8 of the bibliography contains books covering Chapter 14, the conclusions.

The system of reference is as follows: the first number refers to the part of the bibliography; the second to the number of the book in that part; and the third, when given, to the page in the book referred to. Thus 2.5.56 refers to page 56 of the item numbered 5 in the bibliography for Part 2, i.e. Farrington's *Science in Antiquity*.

PART 4

The Birth of Modern Science

Introduction to Part 4

The development of towns, trade, and industry that was gaining moment-um towards the end of the Middle Ages was to prove incompatible with the economy of feudalism. These changes slowly maturing under the surface of the feudal order finally found expression, and in one place after another inaugurated a new order in economy and science. With better techniques, better modes of transport, and more ample markets, the production of commodities for sale steadily increased. The towns where these markets were found had long played a subsidiary, almost parasitic, role in feudal economy; but by the fifteenth century the burghers or bourgeoisie had grown so strong that they were beginning to transform that economy into one in which money payments and not forced services determined the form of production. The triumph of the bourgeoisie, and of the capitalist system of economy which they evolved, took place only after the most severe political, religious, and intellectual struggles. Naturally the process of transformation was slow and uneven; it had begun already in the thirteenth century in Italy, yet it was not until the mid seventeenth century that the bourgeoisie had established their rule even in the most progressive countries of Britain and Holland. Another two hundred years were to elapse before the same class had come to control the whole of Europe.

The same period – 1450–1690 – that saw the development of capitalism as the leading method of production also witnessed that of experiment and calculation as the new method of natural science. The transforma-tion was a complex one; changes in techniques led to science, and science in turn was to lead to new and more rapid changes in technique. This combined technical, economic, and scientific revolution is a unique social phenomenon. Its ultimate importance is even greater than that of the discovery of agriculture, which had made civilization itself possible, because through science it contained in itself the possibilities of indefinite advance.

The problem of the origin of modern science is at last being recognized as one of the major problems of all history. Professor Butterfield [4.1] claims for instance that

the so-called scientific revolution . . . outshines everything since the rise of Christianity and reduces the Renaissance and Reformation to the rank of mere episodes, mere internal displacements within the system of medieval Christendom. . . . There can hardly be a field in which it is of greater moment for us to see . . . the precise operations that underlay a particular historical transition, a particular chapter of intellectual development.

While I disagree profoundly with the analyses he gives, I fully concede the importance of the problem.

The movements of capitalism and science are related, though much too intimately for that relationship to be expressed in simple terms of cause and effect. It can, however, be said that at the beginning of the period the economic factor was dominant. It was the conditions of the rise of capitalism that made that of experimental science possible and necessary. Towards the end of the period the reverse effect was beginning to be felt. The practical successes of science were already contributing to the next great technical advance – the Industrial Revolution. Thus it was in this period that natural science passed its critical point, ensuring its permanent place as part of the productive forces of society. In the longer view of history this fact is far more important than the political or economic events of the time; for capitalism represents but a temporary stage in the economic evolution of society, while science is a permanent acquisition of humanity. If capitalism first made science possible, science in its turn was to make capitalism unnecessary.

In its early stages, however, when capitalism was breaking the bonds of a decaying feudalism, it was vigorous and expansive. The use of the technical devices of the late Middle Ages enabled agriculture, manufacture, and trade to increase and spread over ever larger areas. The material needs of the economic advance led to further developments of techniques, particularly those of mining, warfare, and navigation. These, in turn, led to new problems arising out of the behaviour of new materials and processes, which put a strain on the science of classical times in which such inventions as the compass and gunpowder had had no place. The voyages of discovery showed how limited were the experiences of the Ancients, and strengthened the need to find a new philosophy that could see further and do more.

By the beginning of the seventeenth century a new and enterprising bourgeoisie was able to respond to this stimulus and build up the essentials of experimental science. The new scientists came to be organized, as the merchant adventurers had been, into companies. Before the century was over a small group of able men had been successful in solving the central problems of *mechanics* and *astronomy*. They thus provided more than the science of the Ancients had ever done – practical help where it was needed: in *navigation*. But this was only a slight foretaste; their real triumph lay in the fresh impetus to the scientific study of technique and of Nature, and to the elaboration of the new *experimental* and *mathematical* methods of analysing and solving them, which were to produce their full *fruits* in later centuries. Up to the end of the seventeenth century science had far more to *gain* from its renewed contacts with practical work than it had to *give* in the way of radical improvements in technique.

THE SCIENTIFIC REVOLUTION

The tracing of the development of the new science from the critical period of its birth and early growth to intellectual maturity is the major task of Chapter 7. It is necessary first to show its relation to the new social forces of the Renaissance and Reformation and then to examine how its achievements were to determine the technology and mould the ideas of the Modern age that was to follow. The change in ideas in science in this crucial period was indeed far greater than that in politics and religion, all-important as these seemed at the time. It amounted to a *Scientific Revolution*, in which the whole edifice of intellectual assumptions inherited from the Greeks and canonized by Islamic and Christian theologians alike was overthrown and a radically new system put in its place. A new quantitative, atomic, infinitely extended, and secular world-picture took the place of the old qualitative, continuous, limited, and religious world-picture which the Muslim and Christian schoolmen had inherited from the Greeks. The hierarchical universe of Aristotle gave way before the world machine of Newton. During the transition destructive criticism and constructive synthesis came so close together that it is impossible to draw a line between them.

This substitution was only a symptom of a new orientation towards knowledge. It was changed from being a means of reconciliation of man with the world as it is, was, and ever will be, come doomsday, to one of controlling Nature through the knowledge of its eternal laws. This new attitude was itself a product of the new concern with material wealth and brought about a renewal of interest of the learned in the practice of the trades of the artisan. In this way the Renaissance healed, though only partly, the breach between aristocratic theory and plebeian practice which had been opened with the beginning of class society in early civilization and which had limited the great intellectual capacity of the Greeks.

To understand adequately how modern science began it is necessary to consider both the practical and intellectual aspects of the transformation started in the Renaissance. Writers on the history of science have usually stressed only the latter, and have thus seen the whole transformation either as one from bad to good arguments from self-evident first premises, or as a matter of more careful observation and more correct evaluation of evident facts. That both these explanations are inadequate is shown by their failure to account for the coincidences of the times and places of economic, technical, and scientific advance, and further by the coincidence of subjects of interest to science with those of technical concern to the controlling groups of society.

On the other hand it is also inadequate to consider only these technical interests. Mental attitudes as well as material concerns must be taken into account. The ideological aspects of the struggle of the emerging bourgeoisie impressed themselves on the scientific as well as the religious ideas of these centuries of transition. Indeed, the challenge to ideas that had been accepted for many centuries could only have been made at a time when the whole foundations of society were in question.

Unlike the previous transitions, where, as at the end of the Roman Empire, a new science was built up on the ruins of the old, or where, as at the beginning of the Middle Ages, science was translated from one culture to another, the revolution which gave rise to modern science occurred without any such break in continuity or outside influence. This emphasizes still further the fact that a radically new system of thought was being built up in the new society from elements derived directly from the old, but transformed by the thoughts and actions of the men who were making the revolution. The old feudal culture had been tried and found wanting; it could not survive the conflicts that it had itself engendered. The new bourgeois class which it had thrown up had to find their own new social system and evolve their own new system of ideas. The men of the Renaissance and of the seventeenth century certainly felt they were making a break with the past, however much they might unconsciously owe to it.

In one significant respect the Scientific Revolution differed from earlier changes in that it was made easier, especially at the outset, by the consciousness that it was a return to the ideas of an older, grander, and more philosophical culture. The authority of the ancients could be and was used by such real innovators as Copernicus and Harvey (pp. 406 f.,

437 ff.) to give them a support, not less important than the evidence of the senses. It was a matter not so much of rejecting all authority as of buttressing one against another. The humanist was free to choose, and he could do so for intrinsic reasons. The recovery of at least part of the best mathematical work of classical antiquity, notably that of Apollonius and Archimedes, helped to break the monopoly of Aristotle. And Plato as a mathematician rather than as a theologian could be a source of inspiration. In a sense, and indeed the best sense, the new science came direct from the Ancients; for it was by following their methods that the men of the new age were able to overthrow their ideas and surpass their achievements.

MAJOR PHASES IN THE TRANSFORMATION OF SCIENCE

In order to understand the actual process of the creation of the new science it is convenient to divide the whole period of the Scientific Revolution into three phases which may for convenience be called: those of the Renaissance, 1440–1540; of the Wars of Religion, 1540–1650; and of the Restoration, 1650–90. It must be kept in mind that these are not three contrasting eras but three phases of a single process of transformation from the feudal to the capitalist economy.

In the political sphere the first phase (7.1–7.3) includes the Renaissance, the great navigations, and the Reformation, as well as the wars which ended political freedom in Italy and led to the emergence of Spain as the first great world power.

In the second phase (7.4–7.6) the results of the opening up of America and the East to European trade and piracy began to be felt in a price crisis which shook the whole economy of Europe. It was the age of inconclusive wars of religion in France and Germany. Far more important ultimately for history were the establishment of the Dutch bourgeois Republic at the beginning of the period and of the British bourgeois Commonwealth at its end.

The third phase (7.7–7.9) was one of political compromise. Though governments were monarchical, the big bourgeoisie held the threads of power in all those countries that were progressing economically. The Dutch set the tone of the period, despite the pomp of the Grand Monarque at Versailles. In Britain this phase marked the beginnings of constitutional monarchy and of rapid commercial and industrial development.

The corresponding developments in science were in the first phase a challenge to the whole world-picture which the Middle Ages had adopted from classical times. That challenge found its decisive expression in the rejection by Copernicus of the earth-centred cosmos of Aristotle, and its replacement by a solar system viewed from a turning earth, a planet like the others.

In the second phase the challenge was made good against heavy opposition by Kepler and Galileo and further extended to the human body by Harvey. This was achieved through the use of the new experimental methods, while the first prophets of a new age in science appeared in Bacon and Descartes.

The third phase marked the triumph of the new science, its rapid growth with its spread to new fields, and its first organization into societies. It is the age of Boyle, Hooke, and Huygens, of a new mathematical–mechanical philosophy. The work of many hands and minds ended in Newton's formulation of the *Mathematical Principles of Natural Philosophy*, a foundation on which it was felt that the rest of science could confidently be built. Final ends gave way to mechanical causes, and the hierarchical universe of the Middle Ages was superseded and replaced by another. From now on, independent particles could interact freely, guided by the invisible constitution of Natural Laws. In turn the knowledge of these laws was seen to be the key to the harnessing of the powers of Nature in the service of man. Sublime contemplation had given way to profitable action.

The Scientific Revolution

7.1 The First Phase: The Renaissance 1440–1540

The first phase of the transition from feudalism to capitalism is the period which covers the movements of the Renaissance and the Reformation, though these, with their antecedents and effects, extend over a longer period. The economic pattern of commodity production for a market dominated by money payments had existed in scattered cities since the twelfth century. It first became the prevailing form of economy in the fifteenth century along the strip of country reaching from Italy, through High Germany and the Rhineland, to the Low Countries. Of this area it was only in Italy that the greater cities like Venice, Genoa, Florence, and Milan became politically, as well as economically, independent and were able to build up the brilliant artistic and intellectual civilization of the Renaissance. In Italy this implied no break with the Church, for the Holy See in Rome made a handsome income from the contributions of all Christendom. It was otherwise when the movement spread to Germany and farther. There it led on the one hand to the assertion of independence of religion on a national basis, expressed in the Lutheran Reformation, and on the other to fierce social strife which found expression in the Peasants' War of 1525–6 and the revolt of the Anabaptists of Münster in 1533–5. Similar revolts occurred in Hungary and even in Catholic Spain. Later, when the Reformation spread farther, to the Low Countries, Britain, and France, it was in the still more radical form of Calvinism, rejecting the whole hierarchical Church government and vesting civil as well as ecclesiastical power in the democracy of the elect.

The issue of democracy was, however, not to be raised effectively until the next phase. The political form that was at first to replace the feudal system of graded powers and loyalties was to be the absolute prince, relying for his power on the support of the merchants, and who might even be an ennobled merchant himself, like the Medici. The restoration of monarchy marked an end of the temporal powers of emperor and pope, and with it the whole scheme of the medieval universe. Instead, nation States began to emerge, with ever-shifting alliances and wars between them, resulting in a precarious *balance of power* in which no one could be supreme.

The courts of these kings or princes provided the patronage for the new humanists and scientists, now no longer dependent on the Church. Indeed the position of the intellectuals became very similar to what it had been in the days of the Arabs, when the learned were also the ornaments of princes. The old medieval universities remained, outside Italy, the stronghold of feudal ideas and opposed the new learning. King Francis I of France was obliged in 1530 to found the Collège Royal, now the Collège de France, to provide for the teaching of the humanities which the Sorbonne would not tolerate.

The Renaissance and the Reformation are two aspects of the same movement to change the system of social relations from that based on a fixed hereditary status to one based on buying and selling commodities and labour. The major economic factor that provided the drive for the movement was the rapid extension of trade made possible by a greater available surplus. This surplus was due to the effect of the technical improvements introduced in the latter Middle Ages, particularly those in agriculture and cloth-making.* At the same time the availability of the surplus was enormously increased by improvements in shipping and

101. One of the centres of Renaissance trade and thought was the city state of Venice. The house in this painting by Vittore Carpaccio – active 1490–1526 – shows a rigorous formality and Romanesque style arches, both epitomizing the mood of the times, as does the dress of the courtiers in the foreground.

navigation. Throughout the fifteenth century the main current of trade, still largely of luxury goods, flowed from the East through Venice into Germany to make the fortunes of Augsburg and Nürnberg, and then to the Low Countries and Britain. It was this trade indeed that gave those areas the leading position in wealth and culture.

However, at the end of the century, at the very culminating point of the Renaissance, there was a critical break-through in the old trade pattern, and one in which science played a decisive part. The development of navigation was to short-circuit the old expensive land-based routes to established markets and to open cheap routes to unimagined new markets. The most spectacular result was the discovery of the New World of America, but more immediately important were the Portuguese capture of the Asiatic sea trade and the rapid development of the Baltic lands and Russia. These shifts of trade routes were to alter the whole economic balance of Europe. The trade of Italy and High Germany was cut off at the roots and their political and economic importance was to decline, though their cultural and technical influence was to continue for some time longer. Into their place were to come the maritime countries, first Portugal and Spain, and then for a longer period, because they possessed more basic resources, Holland and Britain.

The profits from overseas trade made possible the first accumulation of fluid capital, that is of capital invested in productive enterprise and not only in land. The greed for more profit led to a rapid development of shipbuilding and navigation, of which the latter was to have a decisive effect on the birth of modern science. With paid soldiers instead of feudal levies, wars could be maintained for longer, but they cost more, hence a demand for bronze and iron, for silver and gold. Mining and metal-working boomed, so did the manufacture of gunpowder and the distillation of strong spirits.

The period as a whole was one of economic expansion. In almost every part of Europe production, not only industrial but also agricultural, increased. There was more grain, more cattle, and more fish. It is difficult to attribute this to any specific technical advance; rather was it the result of the accumulation of innumerable separate improvements, together with the more rapid dissemination of improvements through the new channels of trade. The only radical and important technical advance was the introduction of printing, already discussed for convenience in the preceding chapter (pp. 325 f.). Though this, in itself, was not a method of production, it was one of the most effective ways of disseminating technical advances, as the number of early printed books on such subjects as agriculture, gardening, cooking, and the trades bears witness.

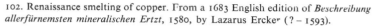

102. Renaissance smelting of copper. From a 1683 English edition of *Beschreibung allerfürnemsten mineralischen Ertzt*, 1580, by Lazarus Ercker (? – 1593).

THE HUMANIST REVOLUTION IN ATTITUDES AND IDEAS

If the Renaissance had only marked a gradual or even rapid improvement of economic conditions it would not occupy the place it does in world history. What gives it its importance in science, art, and politics is that it was a conscious movement, and a revolutionary movement at that. In its intellectual aspect it was the work of a small and conscious minority of scholars and artists. They had set themselves in opposition

103. A sixteenth-century armourer, from a set of 115 woodcuts of arts and trades by Jost Amman (1539–91).

to the whole pattern of medieval life, and they strove to create a new pattern as near as possible to that of classical antiquity. They no longer wished to see the Ancients through the long chain of tradition, through the Arabs and the schoolmen, but directly, by digging up the statues, by reading the texts for themselves. This meant going back to the original Greek and encountering at first hand the thought not only of Plato and Aristotle but also of Democritus and Archimedes.[4.17]

The *humanistic* movement had indeed started in Italy as early as the fourteenth century with Petrarch and Boccaccio. What they appreciated in the classics was beauty of expression and nobility of sentiment, rather than subtleties of logic. In so far as they were philosophical they were Platonist. The humanist movement was to spread to France and northern Europe in the sixteenth century, taking on a more religious flavour. Everywhere it implied a rejection of the specifically feudal ideas of hierarchy and a more secular attitude to society. This did not imply a rejection of religion or even of mysticism, but rather a change of emphasis towards a more personal religion for which the ministrations of the Church were less needed. The cult of the individual, of virtue, in the old Roman sense of manly independence, became the ideal.[4.43]

In Protestant countries the right of private judgement or of special election was proclaimed. Here the humanists, by recovering Greek and Hebrew texts and translating them directly into the vernaculars, were to bring an added weight to the authority of the Bible. Reliance on the literal word of God was to replace deference to the pronouncements of the successors of St Peter. All this fitted an ethical system of a merchant class rejecting the subordination of feudalism. The feudal past was indeed violently repudiated, and with it the architecture which they – the humanists – called Gothic in derision, the philosophy of the schoolmen, the contemplative lives of the monks, the begging of the friars.[4.125] In the end even the Catholic Church itself was forced to reform and accept a break with its medieval past almost as great as that which the Reformers demanded. The doctrine of grace was the Roman equivalent to salvation by faith. The Papacy, which for a century had been in the hands of tolerant humanists, of doubtful morality but great patrons of the arts, was to become almost as rigid as and more intolerant than the most severe Protestant sectary (*n.* p. 461).

PLEASURE, ART, AND MONEY

In Catholic and Protestant lands alike the Renaissance marked a definite and deliberate break with the past. Much of it was inevitably retained, but a new direction was taken and the medieval forms of economy, of building, of art and thought were to vanish for ever, and to be replaced by a new culture, capitalist in its economy, classical in its art and literature, scientific in its approach to Nature.

The Renaissance was a disturbed but hopeful period compared with the despair of the late classical age and the resignation of the ages of faith that followed. There was less concern with the future life and more with the life of the present, a concern which expressed itself in a rapid growth of secular arts, of painting, poetry, and music. In every form of expression there was a new and frank admission of physical enjoyment. The great prophet of this period, Dr François Rabelais (*c.*1490–*c.*1553), chose as the motto for his Abbey of Thelema, the ideal community: 'Do what you like' (p. 1046).[4.125] Ideally, people lived freely and thought dangerously; in fact, few could afford to do so; this new life was expensive and it had to be paid for cash down. Money had become much more important than it had ever been before. As a natural consequence the attitude towards the making of money changed. Any way was good as long as it worked, whether by honest manufacture or trade; by putting forward some new profitable device; by opening a mine; by raiding the foreigners; by lending money at interest. The Church might object, but if it pressed its objections, so much the worse for the Church, as the Reformation showed. Even magic acquired a new interest as a means to wealth and power, as we find in the story of Faust. Indeed natural magic was hardly distinguishable from science.[4.6; 4.20; 4.152]

THE MARRIAGE OF THE CRAFTSMAN TO THE SCHOLAR

Just because they were essential to the making as well as to the spending of money, the technicians and artists were no longer so despised as they had been in classical or medieval times. The arts of ornament and display, painting, sculpture and architecture, flourished and were developed less

104. The manufacture of opaque coloured glass was one of the great arts of ancient Egypt, but by the sixteenth century, clear glass was being produced in large beehive furnaces fired by charcoal. From *De la Pirotechnia* by Vannoccio Biringuccio (1480–1539), published at Venice in 1540.

massively but with far greater originality than in classical times. What was really new, however, was the respect given to the practical arts of spinning, weaving, pottery, glass-making, and, most of all, to the arts that provided for the twin needs of wealth and war – those of the miners and the metal-workers. The techniques of the arts were of more account in the Renaissance than in classical times because they were no longer in the hands of slaves but of free men, and these were not, as they had been in the Middle Ages, far removed socially and economically from the rulers of the new society. In medieval Florence, for instance, the artists had been subordinate members of the major guild of doctors and spice-dealers, *Medici e Speciali*; the sculptors were lower down with the minor guild of the masons and bricklayers.[4.28] By the beginning of the sixteenth century, however, individual painters and sculptors could command the favours of popes and kings, though they often had to press hard to obtain payment for their work.

The enhancement of the status of the craftsman made it possible to renew the link between his traditions and those of the scholars that had been broken almost since the beginning of the early civilizations. Both had a great contribution to make: the craftsman could add to the old techniques of classical antiquity the new devices that had arisen during the Middle Ages; the scholar could contribute the world views, the ideas, and possibly most of all, the logical methods of argument derived from the Greeks by way of Arabic and scholastic philosophy, and the newly evolved methods of computation. The combination of the two approaches took some time to work out, and spread rather gradually at first through the different parts of knowledge and action. But once the constituents had been brought together there was no stopping the combination – it was an explosive one. The intellectual task of the Renaissance was essentially the rediscovery and mastery of the world of art and Nature.

THE WORLD SURVEYED

The Renaissance abounded in great descriptive works covering between them the whole field of human experience. The extent of its interest appears in the achievements of the one man who was himself the epitome of the age – the great universal engineer, scientist, and artist, Leonardo da Vinci. Its two greatest triumphs are the clear statement of the system of the heavens with the sun as centre, the system of Copernicus in his *De Revolutionibus Orbium Coelestium*,[4.128] and the first complete anatomy of the human body pictured in the *De Humani Corporis Fabrica*[4.160] of Vesalius, both published in the same year, 1543. These were the first

ANDREAE VESALII
BRVXELLENSIS, SCHOLAE
medicorum Patauinæ profefforis, de
Humani corporis fabrica
Libri septem.

CVM CAESAREAE
Maieft. Galliarum Regis, ac Senatus Veneti gra-
tia & priuilegio, ut in diplomatis eorundem continetur.

105. Andreas Vesalius (1514–64) is widely acknowledged to be the father of modern
anatomy, and his *De Humani Corporis Fabrica*, published in 1543, contained the
results of his dissections of the human body. This title page, which shows Vesalius
himself beside a dissected body demonstrating its anatomical structure, epitomizes
the high standard of the profusely illustrated volume, one of the most beautiful
scientific books ever published.

106. Biringuccio's *De la Pirotechnia* (see plate 104) contained the first printed practical instructions on various aspects of Renaissance technology. Here wire is being drawn by water power, and at every turn of the crank the metal is pulled through the draw hole.

pictures of how the heavenly spheres or the human body would appear to those who had eyes clear enough to see for themselves, and not through the spectacles of ancient authority. They were put forward and accepted at the outset by a new lay society that was also learning to see and experience for itself. It was only later, when the political conse-quences of the new vision began to be apparent, that authority took fright and tried, too late, to shut it out.

The great works were accompanied by many others in diverse fields of art and Nature which had been neglected by the Ancients. Such for instance was the *Pirotechnia*[4.137] of Biringuccio (1480–1539), describ-ing the metal, glass-working, and chemical industry; and *De Re Metal-lica*[4.23] of Georg Bauer or Agricola (1490–1555), probably the finest technical treatise ever written, for it described not only minerals and metals but also the practice and even the economics of mining. Later there were to appear in such books as those of Gesner (1516–65), Ronde-let (1507–66), and Belon (1517–64) many magnificent descriptions of animals and plants, both of the old and the new worlds.[4.74; 4.32] To these may be added the almost innumerable accounts of the explorations of new lands, including Amerigo Vespucci's *Letters*[4.161] in 1504, which was rather inconsequentially to give the new-found continent its name, and Pigafetta's first account of Magellan's voyage round the world in 1519–22.

The opening phase of the Scientific Revolution was one of description and criticism rather than constructive thought. That was to come later. First must come the exploration of wide horizons and the challenge to old authority. The pursuit of the arts and techniques furnished the positive incentives and the material means for the advancement of the new science. The religious controversies and conflicts shook the framework of orthodoxy, and allowed a few people to try to think for themselves. The new religious attitudes of individual judgement and immediate responsibility stemmed from the same need that was to give rise to science. They were essential preconditions for the triumph of capitalist economy. Before attempting to discuss the position and influence of science in Renaissance life it is first necessary to say something of the influence of the most important factors that affected it in this phase. These are principally those of art and technique, in particular the techniques of engineering and navigation.

7.2 Art, Nature, and Medicine

RENAISSANCE ART

The exaltation of visible and manual art, as against passive and detached contemplation, was the first characteristic of the Renaissance. Painting, sculpture, architecture, and music had, it is true, flourished throughout the Middle Ages. They had been the medium of transference of many of the techniques of classical times, particularly of chemistry and metalworking. They were, however, used as a means to an end, carried out by humble craftsmen or monks in the service of the Church, and to a much lesser extent in that of chivalry.

The social and economic importance of art in the Renaissance was, however, of a different order. Not only was far more spent on art, especially on painting, than in earlier ages, but for the first time the arts began to be valued for themselves. The artists came to serve the new merchant princes wherever they flourished, first in Italy, then in Burgundy, Flanders and High Germany. There was an insatiable demand for more impressive and striking forms to set off the new style of life of the rich.[4.28] With it came the rise in status of the artist and the setting up in most of the cities of Italy of studios which were at the same time universities and laboratories. Art itself, while not ceasing to be traditional, became conscious and scientific. The artists set themselves new problems and found new material and intellectual solutions to them. At no other time in history have the visual arts had such an effect on the development of science, and it is probably no accident that this interest coincided with the very beginning of the most important transformation in the history of science.

PERSPECTIVE AND VISION

The major directions in which the artists helped to found science were in the development of vision and *perspective*, in the interest in Nature and particularly in the *anatomy* of the human body, and in their employment in civil and military engineering. Leonardo da Vinci divided his time between all these interests and though he was the greatest he was by no means the only one to do so.

The first manifesto of the Renaissance art was the *Trattato della Pittura* by Leon Battista Alberti (1404–72) in 1434. He was the son of a wealthy Florentine family exiled for political reasons. Yet he did not scorn to devote himself to art or to learn from manual workers:

> He would learn from all, questioning smiths, builders, shipwrights, and even shoemakers lest any might have some uncommon or secret knowledge of his craft, and often he would feign ignorance in order to discover the excellence of others.[4.24]

He was one of the first advocates of formal perspective – invented by Brunelleschi early in the fifteenth century. The main aim of painting was for Alberti the representation of three-dimensional figures in two dimensions. Accordingly he demanded of all painters a thorough knowledge of geometry, and used optical aids such as the *camera obscura* for landscapes and the rectangular co-ordinates net for plotting the field of vision. The basic metrical concept of three-dimensional space became almost an intuitive commonplace in the Renaissance, due to the realization of this programme by artists like Masaccio, Piero della Francesca and Mantegna.

Leonardo da Vinci was only expressing a prevailing view when he called painting a science. In his treatise on painting published with his *Paragone*[4.165] he states categorically:

> The science of painting deals with all the colours of the surfaces of bodies and with the shapes of the bodies thus enclosed; with their relative nearness and distance; with the degrees of diminutions required as distances gradually increase; moreover, this science is the mother of perspective, that is, of the science of visual rays.

In answer to those who would condemn painting as semi-mechanical, he argues, in flat contradiction to Plato:

107. Perspective was one of the major directions in which the artist began to found science. In the sixteenth and seventeenth centuries, its principles were used with mechanical aids to provide architectural drawings of considerable accuracy. From Robert Fludd, *Utriusque Cosmi . . . Historia*, Oppenheim, 1617–19.

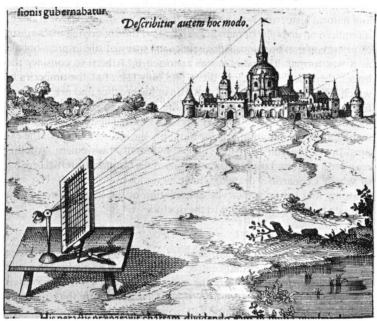

Astronomy and the other sciences also entail manual operations although they have their beginning in the mind, like painting, which arises in the mind of the contemplator but cannot be accomplished without manual operation. The scientific and true principles of painting . . . are understood by the mind alone and entail no manual operation; and they constitute the science of painting which remains in the mind of its contemplators; and from it is then born the actual creation, which is far superior in dignity to the contemplation or science which precedes it.

NATURE AND MAN

The Renaissance saw the triumph of the movement of *realism* in art. Classical, and even more Byzantine, art had been concentrated on ideal forms and the achievement of effect by traditional symbolism. Already in the Middle Ages forms drawn from Nature were beginning to creep in, as foliage and animals, from the sides of the pictures. The Renaissance added the same realism for the central human figures. All this required the most detailed observations of wild Nature – mountains, rocks, trees, flowers, beasts, and birds – and thus laid the foundation of a geology and natural history no longer derived from books and logic. Most of all, it required an *anatomy* of man himself to find the underlying mechanism of gesture and expression. Renaissance art was as little impressionistic as it was formal. The painter was exhorted by Alberti to consider the bones, then the flesh that knit them, and only last of all the draperies in which the figure was clothed. Leonardo in his practice and precept went further. From the representation of the static figure he passed to the moving one, hence to *physiology* and *dynamics*. The representation of men or animals in motion was for him only the means to an end, the expression of the spirit or soul that animates the movement. All this required a profound study of the anatomy of the brain and internal organs, of which Leonardo's drawings have never been surpassed. The new anatomy that led up to Harvey's circulation of the blood owed almost as much to the artists as it did to the doctors.

RENAISSANCE MEDICINE

It is convenient to consider here the great contribution of the Renaissance to the biological studies that centred on medicine. The medical faculties of the Italian universities were the most outstanding exception to their general sterility and obscurantism (p. 297). Particularly in the University of Padua, the medical faculty had acquired the highest prestige and attracted the most brilliant minds. This did not notably help medical practice, for centuries had to pass before enough was known of chemistry and biology to apply science effectively in the battle with disease. It did, however, help enormously in the development of natural science.

The Italian doctors and the large number of foreign students that came there to study medicine were not isolated. They mingled freely with artists, mathematicians, astronomers, and engineers. Indeed many of them followed some of these professions themselves. Copernicus, for instance, was trained and practised as a doctor, besides being an administrator and economist. It was these associations that gave to European, and particularly to Italian, medicine its characteristic descriptive, anatomical, and mechanical bent. The human body was dissected, explored, measured, set down, and explained as an enormously complex machine. The explanation was far too simple; most of what we know now of the function or evolutionary history of the organs was not and could not have been guessed at. Nevertheless a new *anatomy, physiology,* and

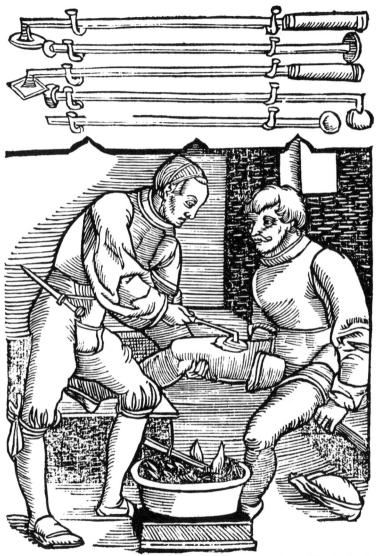

108. Much medical knowledge and practical treatment was learned in the hard world of the military surgeon. Here a wound is being cauterized. From a manual of field surgery, *Veldt Boeck van den Chirugia Scheel-Hans*, by Hans von Gersdorff, Amsterdam, 1593. This Dutch translation contains woodcuts from the Strassburg edition of Gersdorff.

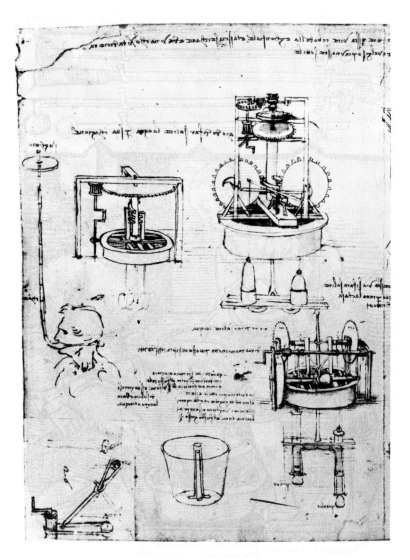

109. Leonardo is remembered for his ingenious mechanical devices or, rather, for the drawings and designs that he made of them. Whether many or any were constructed, we cannot be certain, but his fertile imagination and mechanical knowledge are not in doubt. This page of his sketches is concerned primarily with force pumps of various designs. It also shows an example of Leonardo's 'mirror' writing.

pathology – we owe the last two terms to the great French doctor Jean Fernel (1497–1558) – essentially modern in character, were founded on direct observation and experiment and the hold of classical authority and magical tradition began to be broken.[4.132]

This work found its epitome in the great *De Humani Corporis Fabrica* of Andreas Vesalius, which was the most complete description of all the organs of the body. Yet it still lacked any serious criticism of the classical picture of Galen (pp. 222 f.) and was good anatomy in the service of bad physiology. Nevertheless, the school he founded in Padua in 1537 was to furnish the sequence of anatomists leading up to Harvey.[4.126] Vesalius became physician to the Emperor Charles V. His rival, Francis I of France, had as his surgeon a man who contrasted with Vesalius in many ways, Ambroise Paré (1510–90). He was a real craftsman, unlettered, writing in colloquial French of what he saw with his own eyes and did with his own hands. He revolutionized the treatment of wounds, particularly the gunshot wounds that became so common in the deadly wars of the time.[4.35]

THE ENGINEERS: LEONARDO DA VINCI

The professions of artist, architect, and engineer were not separated in the Renaissance. The artist might be called on by his town or prince, or might offer himself, to cast a statue, build a cathedral, drain a swamp, or besiege a town. The master craftsman had always had to know the properties of materials and the means of handling them. The artist of the Renaissance had to know all that and much more: he had to instil into his work geometry and mechanics in the conscious imitation of antiquity. It was in this field that Leonardo da Vinci, supreme as he was as an artist and a naturalist, showed his greatest ability. In recommending himself to the Duke of Milan, for instance, he cites a number of military devices he can make and adds at the end, 'In painting I can do as well as anyone else.'[4.33.168] His note books show how keenly he studied the operations of metal-workers and engineers and how he himself became the first great master of *mechanics* and *hydraulics*. His greatest attempt, though doomed to failure, was trying to achieve *mechanical flight* – a masterpiece of engineering research, combining observation of birds with the making of models, calculations, and full-scale trials.[4.103; 4.164]

The study of the almost innumerable mechanical devices proposed and drawn by Leonardo, from rolling-mills to mobile canal cutters, brings out another aspect of the tragedy of his genius.[4.10] He could invent machines for almost any purpose and could draw them incomparably well, but hardly any of them, and none of the most important, would have worked even if he could have found enough money to make them. Without a quantitative knowledge of statics and dynamics, and without the use of a prime mover like the steam-engine, the Renaissance engineer could not, in fact, ever advance beyond the limits set by traditional practice. He did not so much affect the development of the machine as impress on the learned world the idea that the operations of Nature could be explained by machinery.

Leonardo da Vinci illustrates in his life and works both the hopes and the failings of the Renaissance.[4.34] Trained as a painter, his many gifts brought him as a youth the patronage of the great at the most brilliant period of Italian art. But he was not satisfied with the practice of painting; he wanted at the same time to understand the underlying nature of what he painted and of the light by which he saw it. Hence his multiple studies on optics, anatomy, animals, plants, and rocks. At the same time he was impressed more and more with the importance of movement and force. It was to realize his ideas in practice that he put himself in the service of the most powerful prince of his time, Ludovico il Moro of Milan, but the shadow of war hung over him and Leonardo could achieve very little there. After the fall of Milan in 1499 Leonardo was forced to become a wanderer – for a while with Cesare Borgia on his campaigns, then in the service of the city of Florence and of the Pope, to die finally as an exile and pensioner of Francis I of France.

All the time he strove to penetrate more deeply into the underlying meaning of Nature and society. In this he was helped by having had no university education and so having less to unlearn; but for the same reason he had neither the systematic approach nor the mathematical skill to follow out his ideas or to convince others of their truth. He left no school and was an inspiration rather than a guide.

RENAISSANCE TECHNOLOGY

The greatest advances of Renaissance technology were in the closely linked fields of mining, metallurgy, and chemistry. The need for metal led to the rapid opening up of mines, first in central Germany and then in America. The German mines were the nurseries of capitalist production. Throughout the Middle Ages mining was largely a series of one-man or small partnership ventures carried out by 'free miners' who were their own prospectors, and who were taxed and protected from minor feudal interference by king or prince.[4.157] With larger-scale mining they came together in *companies* and divided their takings into *shares*. Already in the fifteenth century shares were being taken up by sleeping partners who helped to provide the money for the increasingly costly

110. Pumps of many kinds were used in mining practice. A rag and chain pump from a woodcut in *De Re Metallica* by Agricola (Georg Bauer, 1490–1555). This book, which ran to many editions, was first published at Basle in 1556.

gear. As mines grew deeper, pumping and hauling gear became more essential. Agricola of the *De Re Metallica* was officially a mining doctor in Bleiberg (lead hill) in Saxony, but he also held shares in some of the most profitable mines. The experience gained in power transmission and pumps was the starting point of a new interest in mechanical and hydraulic principles which was to have manifold effects in the Scientific and Industrial Revolutions. With the decay of mining in Germany that set in with the wars of religion, the German miners and metallurgists were dispersed to Spain; to the New World; most important of all, to England, where they provided the technical foundation for her future wealth.

METALLURGY AND CHEMISTRY

The smelting of metals was the real school of chemistry. Extensive mining was bound to bring to light new ores and even new metals like *zinc*, *bismuth* (golden metal), *cobalt* (from kobold, the mine elf), and Kupfernickel (false copper). The ways of separating and handling these had to be found by analogy and corrected by bitter experience; but in doing so a general theory of chemistry, involving oxidations and reductions, distillations and amalgamations, began to take form, at first implicitly. Assaying, to find the yield of an ore in precious metal, is only smelting on a small but definite scale. It became the basis for chemical *experiment* and chemical *analysis* (pp. 620 f.).

The host of new metallic substances could not but have physiological effects, mostly bad, but some good, on those who worked on them. Girls in the mining districts, for instance, used arsenic for improving their complexions. Metallic compounds began to be introduced into medicine on account of their violent effects on the body and to break down the reliance on traditional herbal simples. Particularly decisive was the use of mercury, where ancient herbs proved useless to cure the new and terrible disease of syphilis brought back by Columbus' sailors.

PARACELSUS AND THE DOCTRINE OF SPIRITS

Philippus Aureolus Theophrastus Bombastus von Hohenheim (1493–1541), who called himself Paracelsus to show his superiority to Celsus the great doctor of antiquity, was the intemperate and enthusiastic founder of the new school of iatro-chemists (chemical doctors). He publicly burnt the books of Galen and Avicenna in the market-place of Basle, and in the real protestant spirit proclaimed the supremacy of direct experience over any authority. Though he drew on the old traditions of alchemy transmitted by the Arabs and by Raymond Lull (p. 301), he was able to transform them and change their direction. To the old opposites of *sulphur* and *mercury* he added the neutral *salt*, thus establishing the *tria prima* – rival elements to Aristotle's four (pp. 199 f.) – as the foundation to his *spagyric* art of chemistry, which abandoned the search for gold in favour of the search for health.

Paracelsus' approach to chemistry was frankly animistic. The doctrine of the operation of invisible agencies connected with all self-moving or living activities is one of the oldest of human ideas, probably originating far back in the Old Stone Age. It was associated with the *breath* which came first to every animal at birth and departed with death. The number of words in our language, borrowed from many others, denote the wide ramification of this idea: animal, afflatus, aspiration, ghost, inspiration, psyche, spirit, soul. Air itself was a kind of spirit and its working in bodies, as shown by bubbles, a sign of an active *fermentation*. The crucial process of chemistry, distillation, was essentially a means of capturing the invisible spirits that rose from a boiling liquid. That such spirits were indeed powerful was only too evident from the effect of drinking them (pp. 323 f., 620).

All the operations of the body, according to the Galenic physiology, were performed by several distinct spirits or souls: the vegetable or natural spirit, seated in the liver, presided over the digestion of food; on meeting the vivifying breath in the heart, this became the vital spirit which was spread by the arteries throughout the body; in the ventricles of the brain this in turn was refined into the animal spirit, which, passing through the nerves, gave motion to all the body. Paracelsus, though he rejected Galen, was even freer in his adoption of the concepts of spirits. He pictured spirits – *archaei*, like the little kobolds that haunted the mines – presiding over the various internal organs – stomach, liver, and heart – just at the time when the directing angels were being banished from the heavenly spheres. Nevertheless, owing to the intrinsic complexity of chemistry, it was this intuitive and mythical approach rather than the rational, mechanical one that was to be most successful in advancing chemistry until its revolution in the eighteenth century, and Paracelsus has an undisputed place as the founder of modern chemistry. Even his *archaei* have returned in numbers far greater than he imagined them as the enzymes of modern biochemistry (p. 887).

Metallic ores were not the only minerals that occupied Renaissance chemists. Some, like Bernard Palissy (1510–c.1590), studied earths with the object of finding new glaze for pottery, at a time when European potters were just beginning to catch up with the technical triumphs of the Persian potters. It was still long before they could imitate Chinese porcelain or 'china' as we still call it. Of far greater economic importance was the concern with alum, an essential material in the cloth and leather industries. The possession of alum mines provided ready cash for the Papacy, which founded the first chemical trust, the Societas Aluminum, in 1462.[4,135] Unfortunately papal alum was dear, and the effort to enforce the monopoly by threats of hell-fire was another reason impelling the clothiers of the North to favour the Reformation. In the celebrated indulgences issued by the popes to pay for the building of St Peter's, and which led to Luther's defiance of Rome, we find that among the few crimes for which even the indulgence could obtain no forgiveness was the traffic in alum from rival sources.

Another great chemical development was in the art of distillation, which was so expanded and improved that it underwent no further serious change until well into the eighteenth century. Not only were strong spirits drunk on a large scale in Europe, but they proved to be second only to gunpowder in persuading ignorant savages to give up their lands and even their bodies. By the end of the Renaissance the chemical laboratory, with its furnaces, retorts, stills, and balances, had taken a shape that was to lead without any radical change to the laboratories of today.

7.3 Navigation and Astronomy

VOYAGES AND DISCOVERIES

The technical developments in mining and metals owed little to science, though they gave much to it. It was otherwise with the great voyages that were to open the whole world to European capitalist enterprise. These were the fruit of the first conscious application of astronomical and geographic science to the service of glory and profit. It was natural that the Italian and German cities – Venice, Genoa, and even inland Florence and Nürnberg – with their widespread trade, should take the lead on the theoretical side. There was a resurrection and extension of Greek geography brought up to date by old travellers' reports such as those of Marco Polo and Rubriquis in the thirteenth century and by the results of recent ocean voyages. At the same time the Italians and Germans improved the application of astronomy to navigation, and initiated a drive for astronomical tables accurate and simple enough to be of use to sailors and for maps on which courses could be plotted.

The practical side was primarily the concern of the Portuguese and Spanish sailors, who combined the last effort of the Crusades with a practical eye for sugar plantations, slaves, and gold. Theory and practice met at the Court of Prince Henry the Navigator (1415–60) at Sagres, where Moorish, Jewish, German, and Italian experts discussed new voyages with seasoned Atlantic sea captains. A great revision of the Alphonsine tables (pp. 303 f.) was carried out by Peurbach (1423–61) and by his pupil Regiomontanus (1436–76), who worked in Nürnberg and was later assisted by Albrecht Dürer. In this work they used the old Ptolemaic system but simplified the calculations by means of the trigonometry of Levi ben Gerson (p. 305), thus going back to the Arabs and by-passing the whole medieval mathematical effort. These tables and

111. Columbus's voyage across the Atlantic brought him to a new continent which, however, he did not recognize as such (see page 402). The vessel depicted here is possibly his 100-ton ship the *Santa Maria*. From a woodcut in *Epistola Christofor Columbi, c. 1494.*

methods were of immediate use to the ocean navigators armed with Gerson's cross-staff. In the late fifteenth century the tight monopoly of eastern trade by the Turks made it a tempting idea to break into the Indian Ocean by some other way than the Red Sea. The theorists argued about two possible alternative routes. The most obvious, and one that could be attempted step by step, was to round Africa. This was the way favoured by the Portuguese. It was successfully accomplished in 1488, though India was not reached by Vasco da Gama till 1497. It was by no means certain in advance that it could be done because the land might reach to the Pole, but there were legends that it had been done by the Carthaginians and there might be good pickings on the way.

CHRISTOPHER COLUMBUS AND THE NEW WORLD

The other project which was canvassed among the astronomers and theoretical geographers such as the Florentine, Toscanelli (1397–1482), was to sail westward over the untravelled ocean to find China at the other side of the round world. But to discuss such a hypothesis was a very different thing from making the actual attempt of sailing straight out to sea. In popular imagination anything might happen to such adventurers. They might sail on for ever; they might fall over the edge of the world. The one thing that no one foresaw was that there might be a continent in the way. The man who was willing to make the attempt has always been reckoned the prince of navigators and the most fortunate of explorers, 'A Castilla y a León, Nuevo Mundo dió Colón', though he got little but trouble for it himself. Columbus was very far indeed from being a scientist or having any clear idea of what he was trying to do.[4.81] What he did have was the mystical inspiration that he could, by sailing over the ocean, discover new islands, even Cathaya, or rather, that he was a chosen vessel – Christophoros, the Christ carrier – destined for the discovery of the vision of the apocalypse of 'A new heaven and a new earth'. It was this vision, part religious, part scientific, that gave him the power, as a penniless man, to secure in the end support for his enterprise. It was something that could not have been thought of before and it was difficult enough to do even in the stirring and adventurous fifteenth century. Columbus had to hawk his idea for ten years round the courts of Portugal, Spain, England, and France, being turned down by one expert committee after another. At last it was only by backstairs influence that he got permission to sail with a 100-ton ship and two pinnaces, but holding a contract assuring him of the title, Admiral of the Ocean Sea, and heavy royalties if he should find new land. The contrast between the successive expeditions of the Portuguese round Africa, and of Columbus venturing everything to sail straight across the Atlantic, is typical of that between the technical advance, depending on a steady improvement of tradition, and a scientific one which uses reason to break radically with tradition. For however mystical the internal motives of Columbus himself, the support he received for his voyages was given on the strength of a practical assessment of the return to be expected from the verification of a scientific hypothesis.

Columbus never knew he had discovered a new continent, and it got its name years later from that of the Florentine, Amerigo Vespucci, a learned friend of Leonardo, who was more successful in writing up his discoveries. It was left in the end to the Portuguese, Magellan, in the service of Spain, to complete the proof by showing how to sail round the world. Magellan himself never completed the voyage, being killed in the Philippines. The first man to return to his home after sailing round the world was his Malay slave.

ECONOMIC AND SCIENTIFIC EFFECTS

The economic effects of the great navigations were both immediate and lasting. The short-circuiting, by an open sea route, of the traditional Arab overland and transhipping trade that had been so rewarding to them and to the Turks, brought immense profits to the Portuguese while ruining the Venetians. Later, the exploitation of the mines and the sugar and tobacco plantations of America by means of slave labour snatched from Africa was to bring in a larger and more stable income to Spain and the other colonial powers. Owing to the backwardness of the Spanish economic system this wealth, however, did not stay in the country, for the exploitation both of mines and of trade was in the hands of foreigners, and went to furnish the capital for the industries of Holland and Britain.

The effects on science were also decisive. The success of the early voyages created an enormous demand for shipbuilding and navigation. It brought into being a new class of intelligent, mathematically trained craftsmen for compass, map, and instrument making. This was the beginning of a scientific public, and furnished both a training ground and a livelihood for intelligent youths of all classes. Navigation schools were founded in Portugal, Spain, England, Holland, and France.[4.148] The motion of the stars now had a cash value (p. 427) and astronomy stood in no danger of being neglected, even after astrology had gone out of fashion.

At the same time the twin discoveries of the old and wealthy civilizations of Asia, and the new world of America, with all their strange

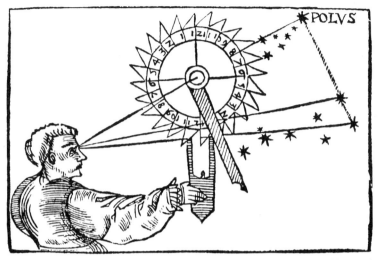

112. It was important for navigators to be able to determine time at night. The usual device was the nocturnal, which was used in conjunction with the Plough and the north pole star Polaris. From Peter Apian, *Cosmographia*, Antwerp, 1539. This book was published posthumously, edited by Apian's pupil Gemma Frisius.

customs and products, made the classical world seem provincial and encouraged men with the knowledge that they had achieved something new which the Ancients had not even been able to think of. The new field, now opened to observation and description, needed new methods to analyse it. The navigations represented indeed as great a break-through in the sphere of the intellect as they did over the sphere of the earth. The initiators of the Renaissance hoped and worked for a new age. By the mid sixteenth century they could feel they had achieved it. The humanist, Jean Fernel, physician to the King of France, and the first man of modern times to measure a degree of the meridian, expresses the new spirit in his *Dialogue* in about 1530. In justifying new ways in medicine he says:

But what if our elders, and those who preceded them, had followed simply the same path as did those before them?... Nay, on the contrary it seems good for philosophers to move to fresh ways and systems; good for them to allow neither the voice of the detractor, nor the weight of ancient culture, nor the fullness of authority, to deter those who would declare their own views. In that way each age produces its own crop of new authors and new arts. This age of ours sees art and science gloriously re-risen, after twelve centuries of swoon. Art and science now equal their ancient splendour, or surpass it. This age need not, in

113. Nicholas Copernicus (1473–1543). This contemporary woodcut is probably, with two other contemporary woodcuts, the most authentic of all the likenesses of Copernicus and shows him holding a lily-of-the-valley in one hand.

any respect, despise itself, and sigh for the knowledge of the Ancients. . . . Our age today is doing things of which antiquity did not dream. . . . Ocean has been crossed by the prowess of our navigators, and new islands found. The far recesses of India lie revealed. The continent of the West, the so-called New World, unknown to our forefathers, has in great part become known. In all this, and in what pertains to astronomy, Plato, Aristotle, and the old philosophers made progress, and Ptolemy added a great deal more. Yet, were one of them to return today, he would find geography changed past recognition. A new globe has been given us by the navigators of our time.[4.132.17]

THE COPERNICAN REVOLUTION

It is no accident that it was in the field of astronomy, so closely related to that of geography, that was to come the first and in some ways the most important break in the whole ancient system of thought. This was the clear and detailed exposition by Copernicus of the rotation of the earth on its axis and its motion around a fixed sun. Descriptive astronomy was the only science at that time which had accumulated enough observations and developed mathematical methods accurate enough to permit hypotheses to be set out clearly and tested numerically. Also, as we have seen, it was a centre of renewed interest both for its old astrological and its new navigational use. These in themselves might well not have led to any radical advance. Professional astronomers like Peurbach (1423–61) and Regiomontanus (1436–76) found the old methods, with minor improvements, good enough for them. Nevertheless it is to them and the Renaissance spirit which led them to seek for Greek originals that we owe the new astronomy. Peurbach was in the service of Cardinal Bessarion (*c.* 1400–1472), the Byzantine humanist, and was engaged by the Pope in the reform of the calendar.

What Copernicus added was the new critical spirit, an appreciation of aesthetic form and the inspiration of newly edited texts which could also be used to balance one ancient authority against another. For, as we have seen, the idea of the rotation of the earth was by no means a new one. It goes back to the very foundation of Greek astronomy and was stated in so many words by Aristarchus in the third century B.C. (pp. 216 f.). It had always remained as an alternative – though paradoxically absurd – view of the motion of the stars; for it was self-evident that the earth did not move, while the sun, moon, and stars could be seen to do so. Courage as well as science would be needed to upset the common-sense view. The man who was to dare to do this, for all his retiring nature, had plenty of courage and, as a Renaissance humanist, had all the incentives to achieve this decisive break with the past.[4.10; 4.11]

Nicholas Copernicus was born at Torun in Poland in 1473, was educated at Bologna for astronomy, at Padua for medicine, and at Ferrara for law, and spent most of his life as canon of Frauenburg. As this cathedral town was situated in the disputed territory between the Teutonic knights and the kingdom of Poland, he had much to do with war and administration; but his main interest was always astronomical, and he devoted the whole of his private life to the effort to find a more rational picture of the heavens, which he set out in final form in his book *On the*

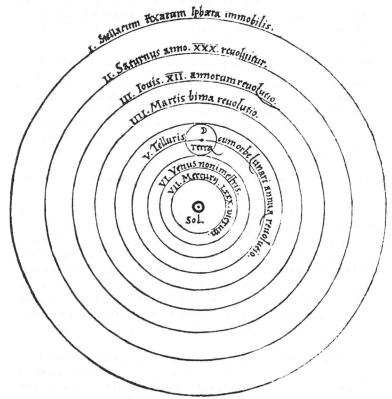

114. The heliocentric system of Copernicus. This apparently prosaic woodcut laid the revolutionary doctrine of a moving Earth before Renaissance culture. The universe still consisted of spheres, but its centre point was no longer the Earth. A new philosophical outlook was to follow, brought about by man's dethronement from the centre of the universe. From Nicholas Copernicus, *De Revolutionibus Orbium Coelestium*, Nuremberg, 1543.

Revolution of the Celestial Orbs which was printed only in the very year of his death in 1543. In it he postulated a system of spheres centred round the sun rather than the earth, assuming the rotation of the earth and showing in detail how this could account for all astronomical observations.* His reasons for this revolutionary change were essentially philosophic and aesthetic.[4.1] Speaking of his sun-centred system and its implication of the almost infinite distance of the stars he writes:

> I think it is easier to believe this than to confuse the issue by assuming a vast number of Spheres, which those who keep Earth at the centre must do. We thus rather follow Nature, who producing nothing vain or superfluous, often prefers to endow one cause with many effects.[4.128.19]

And then after describing the planetary orbs one after another he ends:

> In the middle of all sits Sun enthroned. In this most beautiful temple could we place this luminary in any better position from which he can illuminate the whole at once? He is rightly called the Lamp, the Mind, the Ruler of the Universe; Hermes Trismegistus names him the visible God, Sophocles' Electra calls him the All-seeing. So the Sun sits as upon a royal throne ruling his children the planets which circle round him. The Earth has the Moon at her service. As Aristotle says, in his *de Animalibus*, the Moon has the closest relationship with the Earth. Meanwhile the Earth conceives by the Sun, and becomes pregnant with an annual rebirth.

Here also we see both a return to a most ancient, indeed a magical, view of the universe and an exaltation of the central monarchy, *le Roi Soleil*.

The presentation of the *solar system* took some time to have any effect. A few astronomers appreciated it as a means of improving calculations. The Prussian tables were prepared in 1551 on the basis of the Copernican system, but few believed it was really true. Besides being repugnant to common sense, there were many objections that affected the learned, particularly as to how the earth could go round without producing a mighty wind or deflecting the fall of shot. These were only finally removed by Galileo (pp. 427 ff.).

Nevertheless the mere idea of an open universe, with the earth but a small part of it, was shattering to the old image of closed concentric crystalline spheres, divinely created and maintained in motion. If there were new worlds on the earth, might there not also be in the sky? This was the heresy for which Bruno was to die.

THE ACHIEVEMENT OF THE RENAISSANCE

The first phase of the Scientific Revolution was mainly a destructive one in the field of ideas, though illuminated by the one brilliant constructive

115. Building construction in the sixteenth century differed little from that of previous ages. Block and tackle, winches, and other devices were used to provide mechanical advantage. From Jacques Besson, *Theatre des Instrumens Mathematiques et Mechaniques*, Lyons, 1593.

hypothesis of Copernicus. Not only in astronomy but also in other fields of interest – in anatomy, in chemistry – the old ways of thought were proving inadequate and unsatisfying. The men of the Renaissance, even if they had found the solutions to few of the problems they had raised, had at least cleared the ground for the solution of the remainder in the great struggle of ideas of the succeeding century.

In the use of science, by contrast, the Renaissance marked an era of decisive achievement. The scientific effort of the early Middle Ages had faded away, largely, as has been suggested, because no practical use could be found for it. The achievements of the Renaissance navigators did provide just what was necessary – a secure and growing field of application. And this field required astronomy and navigation, just those parts of science best preserved from classical times and most actively maintained in the service of astrology and calendar-making. Further supports were to be provided for the science of mechanics in the development of machines, and for that of dynamics in the development of gunnery. From now on science was secure; it had become a necessity for the most vital, active, and profitable of enterprises – for trade and war. Later it could extend its services to manufacture, agriculture, and even medicine. The overall importance of the Renaissance was that it marked the first break-away from the economy, the politics, and the ideas of the feudal Middle Ages. Most of the constructive work had still to be done, but there was to be no turning back. Science was beginning to make its mark on history.[4.2; 4.9; 4.76; 4.130]

7.4 The Second Phase:
Science During the First Bourgeois Revolutions 1540–1650

The period roughly from 1540 to 1650 has no convenient name in history. It has been called the Counter-Renaissance,[4.6] but this would indicate a far greater degree of reaction to the earlier phase than actually took place. It includes the Counter-Reformation, with the Baroque style that was its visible expression, the Wars of Religion that raged in turn in France (1560–98), in the Low Countries (1572–1609), and in Germany (1618–48), and the establishment of the States General of Holland in 1576 and the Commonwealth of England in 1649. Of these events it was the last two that were to have the greatest ultimate significance. They point to the political triumph of the new bourgeoisie in the two countries in which was concentrated the bulk of the world's trade and manufacture.

In science the period includes the first great triumphs of the new observational, experimental approach. It opens fresh from the first exposition of the solar system by Copernicus and closes with its firm establishment – despite the condemnation of the Church – through the work of Galileo. It includes in its scope Gilbert's description in 1600 of the earth as a magnet and Harvey's discovery in 1628 of the circulation of the blood. It witnesses the first use of the two great extenders of visible Nature, the telescope and the microscope.

Economically the century was dominated by the cumulative effects of the navigations, which by then involved a trade comparable with the old internal trade of Europe. It was specially marked by the great increase in prices brought about by the influx of American silver. The breakdown of feudal land-holding in western Europe, especially in Holland and England, had thrown on the market landless people, and, at the same time, the real wages of hired workers were seriously depressed. This had the effect of lowering the cost of products in a period of rising prices and increasing markets, and at the same time of providing an abundant labour force for manufacturers. The result was an unprecedented increase in the wealth of those traders and manufacturers who were on the new oceanic trade routes and could draw on new resources and supply new markets. [4.5; 4.14] The combined effects of the change in trade routes and wars were to ruin the economy of Germany, which had been the most progressive part of Europe in the early sixteenth century.

The loss at the old centre was more than made up on the periphery. The new economic centre of Europe, and by then indeed of the world, was in the lands round the North Sea, first Holland, then England and northern France. There, unlike the other maritime countries Spain and Portugal, where conditions remained feudal, manufacture could be combined with trade. The German and Italian craftsmen emigrated and rapidly spread the technical and artistic achievements of the Renaissance to the now dominant northern nations. At the same time the need for corn to feed the growing populations of Holland and England, and for flax, timber, pitch, and iron for their shipping, stimulated the economic development of the Baltic countries, where Denmark, Sweden, Poland, and Russia in turn began to emerge as independent powers.

The moving spirits and the main beneficiaries of this second phase of the economic revolution were the merchants of Holland and England, who were backed by flourishing agriculture and fisheries. Wealth brought political power to the bourgeoisie, but this did not come easily. It took years of struggle and open warfare before the kings, first of Spain and then of England, were forced to realize that they could no longer hold their wealthy Dutch or English subjects under feudal conditions that hindered them in the pursuit of profit. The ostensible reasons for this struggle were religious, and this had at least the justification that the political and economic convictions and practice of the new bourgeoisie were more naturally expressed in Calvinist than in Catholic or even in Lutheran terms.[4.147]

THE ADVANCE OF TECHNOLOGY

Technically the century was one of steady advance in scale and perform-ance, without any of the revolutionary innovations that characterized the centuries before and after. Agriculture was still the predominant occupa-tion and woollen cloth-making the major industry. Nevertheless change was in the air. Shipbuilding improved with experience and with it naviga-tion. The increase in trade and the lowering of transport costs led to a much wider distribution of wealth among the bourgeoisie. Rare luxuries, such as silks and glass, became commodities, while new products from east and west, such as cottons, porcelain, cocoa, and tobacco, began to come into European markets. Painting, in the Flemish and Dutch schools, began to leave the service of religion and the exaltation of nobil-ity to portray ordinary people eating, drinking, and making merry. It is in this time that the Dutch set the standard of bourgeois *comfort* in town and country houses, and invested good money in gardens and fields (p. 474).

THE BLAST-FURNACE AND CAST IRON

An ultimately more momentous change was occurring almost unper-ceived in the methods of production of less spectacular goods, particu-larly of iron. It was in this period that the transformation in iron metal-lurgy that had been maturing in Europe since the fourteenth century first began to have a decisive effect. Cast iron had been known in China since the first century B.C. (p. 147), but its appearance in Europe seems to have been quite independent. Its production is typical of a crucial change brought about by a mere increase in the scale of operation. For 3,000 years iron had been made by low-temperature reduction with char-coal in small bloomery furnaces, leaving the iron in a pasty mass (p. 147). Throughout the Middle Ages these furnaces came to be made larger and the blast for them was provided by bellows, ultimately driven by water-power. Occasionally the temperature was high enough to melt the iron and turn the malleable 'bloom' into an intractable 'bear'.[4.144] Then, first in the Rhineland in the fourteenth century, came the idea of running off the iron on to the floor in front of the furnace into a hollow which soon became the 'sow' with its litter of 'pig' iron. This pig iron was at first difficult to refine and improvement was slow; but, as the knowledge of the process spread, bloomeries gave way to the new *blast-furnaces*, and by the end of the sixteenth century iron began to be poured out by the ton instead of being beaten out by the hundredweight.[5.2]

The limitation that dear iron had imposed on all techniques was rapidly removed, but a new bottle-neck appeared, caused by the shortage of the wood charcoal needed to smelt the larger quantities of iron. Old-established iron regions, like the Weald of Sussex, lost their predominance, which passed to Sweden and Russia with their large supplies of timber. Iron, indeed, was a major factor which, through trade and war, brought them into world economy. Cast iron was used first and foremost for weapons, especially cannon, once the bronze bell-founders' art could be applied to it. England early acquired a reputation for good cannon and they were marketed on strictly business principles. The guns of the galleons of the most Catholic King of Spain and those of the infidel Bey of Algiers were as likely as not to have been cast in Sussex.[4.144]

THE USE OF COAL

The shortage of wood for iron-smelting was only one reason among many for the acute timber crises that affected Holland and England in the late sixteenth century. General mercantile prosperity raised the demand for timber – for ships and houses, for firewood, for salt-makers, soap-boilers, maltsters, as well as for domestic uses – far beyond the capacity of the local forests. Some could be imported, but a remedy lay ready to hand in the pit coal that had been worked from open seams in Northumbria and Scotland since Roman times, and had already found a distant market as sea coal in London and even on the Continent in the Middle Ages. It was pretty filthy stuff, but it came to be burnt by the citizens for fuel despite all laws prohibiting its use.

As the price of firewood soared in the sixteenth century, more and more uses were found for coal, and its production increased rapidly. In the seventy years from 1564 to 1634 the annual shipments of coal from Newcastle rose fourteenfold to nearly half a million tons.[4.112] Correspondingly more technical effort was put into mining it from deeper and thus more easily flooded pits. This led to the use of devices largely taken over from the metal mines of Europe – improved pumps and the wooden *rail way* for running the trucks out of the mines. Coal was indeed to solve the recurrent fuel crises that in the past had driven civilization

116. The first illustration of a railway, showing a wooden rail for mining trucks. From Sebastian Münster (1489–1552), *Cosmographia Universalis*, Basle, 1544.

further and further to uncut backwoods. From then on the centre of industry, and with it the centre of civilization, was to move towards the coalfields, where it was to be fixed for another 400 years at least. It was this, rather than any other factor, that was to lead to the industrial predominance of Britain. As that shrewd observer Daniel Defoe put it in his description of the West Riding of Yorkshire:

> . . . such has been the bounty of Nature to this otherwise frightful country, that two things essential to the business, as well as to the ease of the people are found here, and that in a situation which I never saw the like of in any part of England; and, I believe, the like is not to be seen so contrived in any part of the world; I mean coals and running water upon the tops of the highest hills: This seems to have been directed by the wise hand of Providence for the very purpose which is now served by it, namely, the manufactures, which otherwise could not be carried on; neither indeed could one fifth part of the inhabitants be supported without them, for the land could not maintain them.

Neither in the technical innovations involved nor in the use of science can the industrial upsurge of the late sixteenth and early seventeenth centuries, which has been called the first Industrial Revolution,[4.14] be compared with the great Industrial Revolution of the eighteenth century. Nevertheless, we can see now that it was its essential prelude. Before it was conceivable or possible to change from a wood and water-power technology to an iron and coal one, the change had to be demonstrably necessary. It was the pressure of the demand of the first Industrial Revolution on the limited resources which had sufficed for the feudal economy of the Middle Ages that forced the search for new resources and new techniques.

THE PROJECTORS: SIMON STURTEVANT

It was also that same pressure that finally altered the attitude towards novelty. Once profit was legitimate and novel methods could promise riches, novelty was to be embraced rather than shunned. This was the shop that sold the 'new thinking cap' to which Professor Butterfield attributes the birth of modern science.[4.1] The late sixteenth and early seventeenth centuries saw the first of the race of *projectors*, later called *inventors*. These did not merely talk, as Roger Bacon had done, of wonderful new machines but offered to make them, for a consideration, and sometimes even did so.[4.149]

Hic per uirgam diuinatoriam quæritur latens metallum, & ex puteis in pro/
fundum actis educitur, qui est primus Metallicorum labor.

Virgula diuina

Glück rüt

Haspel

Instrumëtum Tractorium

Grubener

Bürman

zer setzer

Hatwer

Einde mirabili quoq; artificio inclusum in specubus terræ aere recenti aura fol/
liculis & uentilabris ad hoc paratis subinde innouare contendunt, ne metallarij
in illa Plutonis officina aere impuro suffocentur. Habent comites Mansfel
denses mineras in sectilibus seu fissilibus lapidibus, quos Schiffer Germani uo
cant, quales mineras in uniuerso mundo extare non putant. Nã sectiles illi lapides igne ad
usti & macerati reddunt cuprũ, ex quo immensa uis trahitur argẽti. Nec exhauriri potest
fundus illius terræ his uenis. Nam ubicunq; in terra illa fossioni intenditur, inueniuntur se
ctiles illi lapides cupro prægnantes. Fossores, qui lapides illos eruunt seu excindunt, obli/
quum nancifcuntur collum, uocanturq́ue Germanicè Krumhelss, id est, obliqui seu toruí
collarij. Quum enim petrarum illarum uenæ & meatus planum habeant situm, necesse est
ut fossores iacendo intendant labori, unde fit, quod colla illorum sic retorqueantur, ut pla
ne inepti reddantur ad aliarum minerarum labores. Magnum licet ibi in natura uidere
miraculum. Est in ea regione lacus quidam magnus, longus & latus, & qualia ille uiua ani
malia

Such a man was Cornelius Drebbel (1572–1634), who built a sub-marine which he showed in the Thames, but made a more profitable venture in introducing a scarlet dye. Such also was the forgotten and tragic figure of Simon Sturtevant, an eccentric clergyman who aimed higher – at nothing less than 'the working, melting and effecting of Iron, Steele and other mettles with Sea Coale or Pit coale; the principall end of such invention is that the woods and timber of our country might be saved' (from the preamble of his Patent *The Treatise of Metallica*, 1612,[4.145]). What Sturtevant's secret was, or how he came on it, we may never know. The problem he set himself was not solved in practice for another hundred years (p. 596), but he has left us a most precious account, in many ways unsurpassed, of the technical and economic aspects of invention, written before the dawn of the new industrial age. Sturtevant begins with 'Heuretica – the Art of inventions, teaching how to find the new and judge the old'. This he further divides into an 'Organick' part, comprising the fixed capital, and a 'Technick' part, comprising the skill of the 'Artizands'. In his analysis of the processes of invention he distinguishes drawings, models (superficiall and real Moddles), working models, prototypes (the Protoplast), and finally the 'Grand Mechanick', or full-scale production 'set up after the form and type of the Protoplast in greatness, or with some profitable additions which later experience has taught'. He was fully alive to development costs and the criteria of profitability, and had clear ideas as to the means of raising capital. Why then did he fail completely? It was not technical incapacity – he had already proved his worth by inventing pressed clay ware, which we use to this day. The reasons appear in the sequel to arise from the conditions of the time, which were quite unsuited for the kind of capitalist enterprise he foresaw with such surprising clarity.

Sturtevant estimated the annual yield of the iron monopoly at £330,000. He accordingly divided his enterprise into thirty-three shares; of these the King, Princes, and the favourite Carr received eighteen, Sturtevant himself received one, and the remaining fourteen were to be distributed among 'those who shall adventure, joine or assist the work'. What with the court rake-off it is not surprising that it all came to nothing. Two of the adventurers stole the patent from Sturtevant, got him outlawed, and then failed to work the process themselves, since the original patent is a model of obscurity when it comes to particulars.*

Modern industry could not arise from feudal conditions, or even from the prerogative of a Renaissance prince who, lavish in his expenses, was always short of money and always being cheated. The real technical advance was made by the small men building up their capital out of their profits. This they could achieve only in the next century, when the privileges of kings, nobles, and corporations had been swept away (p. 596).

THE NEW EXPERIMENTAL PHILOSOPHERS

It was in this atmosphere that the new, half-awakened science of Europe was to grow to maturity. Despite widespread privilege and corruption it was not by any means an unfavourable one. Even the movement of

117. A contemporary pencil sketch of Tycho Brahe (1546–1601), the doyen of Renaissance observational astronomers.

the Counter-Reformation, which successfully checked and turned back the advance of Protestantism in Europe, had not the same effect on science. The Jesuits who directed it had the intelligence to realize that they were more likely to win souls by fostering science than by blindly opposing it. They accordingly entered fully into the scientific movement, particularly the new astronomy, and were even the agents for spreading it and setting up observatories in India, China, and Japan. At the same time they acted as watchdogs inside science to guard against any damaging effect it might have on true religion, and thus unintentionally gave an advantage to scientists in Protestant countries out of their control.

The fifteenth-century concentration of science in Italy was replaced by a wide diffusion over Europe, though Italian intellectual pre-eminence was for some time to outlast political and economic decadence; for Italy, the first of the countries of western Europe to break away from the feudal tradition, remained the centre of European culture long after she had lost her political and economic importance. That culture was a well-balanced one, since in Italy, at first alone in Europe, the universities had largely been won for the new learning. The professors were moreover courtiers, and were thus able to combine practical knowledge of the world and full acquaintance and contact with scholastic tradition. From whatever country new scientists came, whether from Poland, England, or France, it was in Italy that they acquired their knowledge and it was in Italy that they did much of their best work.

The new experimental philosophers, or scientists as we would now call them (p. 32), no longer formed part of the intense city life of the Renaissance; they appeared more as individual members of the new bourgeoisie, largely lawyers, like Vieta, Fermat, Bacon; doctors – Copernicus, Gilbert, Harvey; a few minor nobles – Tycho Brahe, Descartes, von Guericke and van Helmont; churchmen, like Mersenne and Gassendi; and even one or two brilliant recruits from the lower orders, like Kepler. In history they are made to figure as being isolated; but in reality they were, because of their very small numbers, always in far easier and quicker contact with each other than scientists of today, with their vast numbers and the pressure, publication delays, and increasing military and political restrictions to which they are subjected.

SCIENTIFIC EDUCATION : GRESHAM COLLEGE

In Holland and in England there was even the beginning of scientific education, with a decided bent for navigation, in imitation of the Spanish and Portuguese schools of the first phase. The Flemings, Gemma Frisius (1508–55) and Gerard Mercator (1512–94), had shown the way in making accurate navigational charts. They were closely followed by English geographers, of whom the first, John Dee (1527–1608), though best known as an astrologer, was the friend and adviser of many of the great Elizabethan sailors and may rightly be claimed as the first British scientist of the new age. The first institute for teaching the new science in England was Gresham College, established in 1579 by the will of Sir Thomas Gresham (1519–79), one of the great London merchants, financial agent to the Crown, and founder of the Royal Exchange. He personified the union between merchant capital and the new science. Unlike the Collège de France of the earlier generation (p.380), Gresham College was no mere humanist institution. Lectures were to be in English as well as Latin. Of its seven professors, two were appointed for the sciences of geometry and astronomy, and the latter was urged to lecture on instruments of navigation 'for the capacities of mariners'.[4.79] Gresham College was to be for over a century the scientific centre of England and was to house the Royal Society, which at first met in its rooms.[4.83]

118. Gresham College, from an engraving by George Vertue (1684–1765), probably the best known of all English engravers. From John Ward, *Lives of the Professors of Gresham College*, London, 1740.

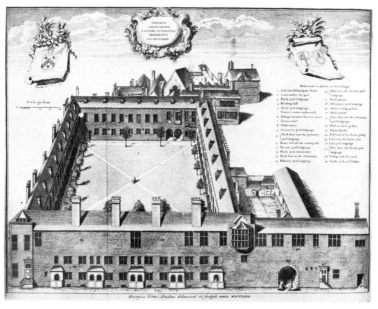

Most of the scientists of the period took for granted, what had been heresy in classical and medieval times, that science was primarily concerned with Nature and the arts and that it was its business to be useful. Most of them at one time or another were in State service and tried to justify their employment by practical inventions in peace and war. Their originality and individualism were only apparent. In most of their thought they necessarily relied on the same traditions, used the same methods, and were drawn to the same problems. These problems were limited in number compared with either the qualitative Renaissance universalism or the systematic search of Nature of the succeeding phase of organized science. The main questions asked were those concerned with the working of the heavens, leading to the use of astronomy in navigation, with the movements of projectiles and machines, and with the gross mechanism of the human body. Their programme was no longer purely negative, as in the first phase of the Renaissance; they set out not so much to destroy the systems of Aristotle and Galen as to provide workable alternatives. In this they succeeded beyond all expectation, though the final synthesis was to be reserved for the age of Newton.

7.5 The Justification of the Solar System

The implications of the Copernican revolution took some time to sink in. It was welcomed most readily by professional astronomers because of its simplicity and as a means, though still far from an accurate one, of improving astronomical tables. Next came those who found in it a convincing illustration of the stupidity of the old, medieval, Aristotelian world view or who were inspired by the vision of an infinite universe which it opened up. Of these the most famous was Giordano Bruno (1548–1600).[4.136] Born in Nola near Naples, of fiery temperament and penetrating imagination, he soon quarrelled with the monastic order which he had joined and led a wandering life throughout Europe, disputing and publishing books and pamphlets in which he mingled Lullian mysticism with the idea of the plurality of worlds. His ability was such that he impressed magnates and scientists alike, but his sharp tongue made him more enemies than friends and he was kept always on the move. At last, venturing incautiously into Venice in 1592, he was betrayed and handed over to the Roman Inquisition, who burnt him to death eight years later for heresy. He was a martyr not so much to science as to freedom of thought, for he made neither experiments nor observations, but insisted to the end on his right to draw what conclusions he chose from the facts of science.

Bruno made people think and argue about the Copernican theory. For all the Catholics that his execution frightened, as many Protestants must have been encouraged. More solid arguments were needed, however, before the Copernican theory could be established and profitably used. What the theory lacked in its first form was an accurate description of the orbits of the planets, which was to be the work of astronomers to provide, and also convincing arguments to justify the imperceptibility of the motion of the earth, a task which implied the creation of a new science of dynamics.

URANIBORG AND TYCHO BRAHE

The first task was carried out by two remarkable men, Tycho Brahe (1546–1601) and his assistant, Johannes Kepler (1571–1630). Tycho Brahe, himself a Danish nobleman, was able to use enough influence with King Frederick II to build in 1576 the first really scientific institute of the modern world – Uraniborg – on the island of Hveen in the Sound from whose tolls Denmark drew most of its wealth. There, with specially made apparatus, he collected a series of exact observations on the positions of stars and planets that made everything that had been done before obsolete. He was influenced by Copernicus' work, but he preferred a system of his own in which the sun turned round the earth but the planets turned round the sun, which is, of course, the Copernican system relative to a motionless earth. In fact he chose the system that best fitted the observations without worrying about its physical absurdity. He had actually, without insisting on it, already shattered the Aristotelian system by demonstrating that the New Star of 1572 lay in the sphere of fixed stars, where by definition no change could take place. Tycho lived in a transitional time for astronomy, just when the old need for astronomical data, almost exclusively for astrological purposes and consequently subsidized only by princes, was giving way to a new need for more exact astronomical data for the use of navigators.

KEPLER

Tycho's results became infinitely more valuable for the progress of science when they were worked over by Kepler. He was the son of poor parents and lived a life of continual struggles and frustrations, partly due to his own strange character. He was the first great Protestant scientist, though

119. Kepler's great computational results, based on Tycho Brahe's observations of the planets, were enshrined in his *Tabulae Rudolphinae*, dedicated to his patron Rudolph II of Austria and Bohemia, and published in 1627. They were computed on the basis of a heliocentric universe using for the planets elliptical orbits which had been discovered by Kepler himself, and their accuracy was such that they were in use for a hundred years. The frontispiece of the *Tabulae* is shown here. On the left is Hipparchus, next to him Copernicus, then Tycho, and finally Ptolemy. The figure in the background represents no notable astronomer. In the centre panel of the base is a map of the island of Hveen where Tycho's original observatory stood, and in the panel to its left sits Kepler himself. It was he who designed this elaborate frontispiece.

120. Kepler's first attempt to build a synthesis of planetary motion was in his method of utilizing the concept of celestial spheres, and computing their distance on the assumption that one of the five regular geometrical solids could be inserted between each. This was first published in 1596 in his *Mysterium Cosmographicum* and brought him to the attention of Tycho Brahe. From Johannes Kepler, *Harmonices Mundi*, Linz, 1619, which was a continuation of Kepler's *Mysterium Cosmographicum.*

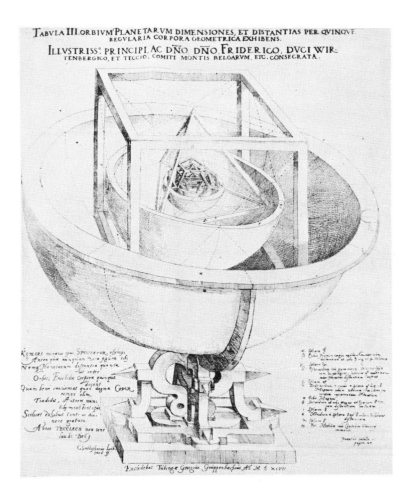

he worked for most of his life in Catholic lands. He combined in a most unusual way a fantastic imagination, deeply tinged with number magic, with a scrupulous integrity in the accuracy of his measurements and calculations. The major drive behind his work was a mystical desire to penetrate the secrets of the universe, as witness the title of his first work, *Mysterium Cosmographicum*.[4·99] But he had to live, and, as he said, 'God provides for every animal his means of sustenance – for astronomers He has provided astrology.' He assisted Tycho in his last years in the crazy alchemical–astrological institute that the Emperor Rudolph II had set up in Prague. The presence of active and subsidized scientific research in sixteenth-century Poland, Denmark, and Bohemia was in itself a sign of the new economic development that these countries, on the fringe of feudal Europe, were then undergoing.

There Kepler tried to find the best way of representing planetary motions by a single curve. Copernicus had still stuck to circles and epicycles, but not only were these clumsy but they could not be made to fit

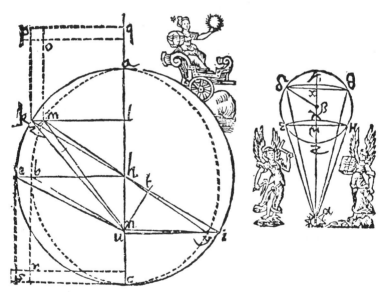

121. The discovery that the planets orbited in ellipses rather than in circles, as had been believed previously without question, was an important step forward. Kepler first discovered such paths from an analysis of Tycho Brahe's meticulous observations of Mars. From Johannes Kepler, *Astronomia Nova . . . de Motibus Stellae Martis*, 1609.

the new accurate observations. Kepler found, after many failures, that the only explanation of the observed movement of the planet Mars was that its orbit was an ellipse with the sun as focus. The idea of elliptical orbits was not completely new; it had been suggested by Arzachel (1029–87) of Toledo in the eleventh century, but on quite inadequate data. Kepler succeeded because he came at a time when the data were exact enough to show that no circle or combination of circles would do, and not so late for them to be so exact that it was apparent that the orbits were not true ellipses but more complicated curves, which were to be explained only by Einstein.

The hypothesis of elliptical orbits, and the two other laws by which Kepler explained the speed of a planet in its orbit, not only removed the main astronomical objection to the hypothesis of Copernicus, but they also struck a mortal blow at the Pythagorean–Platonic view of the necessity of the heavens showing perfect – that is, circular – motions only, which even Copernicus had retained.* These purely astronomical calculations of Kepler were not, however, the decisive element in producing the great revolution in men's minds leading to an altogether new view of the universe, though they were to be the observational basis of a quantitative, dynamical explanation which Newton was later to work out (p. 475).

THE TELESCOPE

The step that was to prove decisive in securing the acceptance of the new view of the heavens was not to be any further extension of astronomical calculation, appreciated only by experts, but a direct physical means available to all of bringing the heavens down to earth so that sun, moon, and stars could be more closely examined; in other words the invention of a telescope or far-seer.

The *telescope* was probably not itself a creation of science: it appears rather obscurely in Holland as a by-product of the manufacture of spectacles. Legend has it that it was about the year 1600 when some child in Lippershey's shop first looked through one lens at another in the window and noticed that it made things outside seem nearer. The fact that no scientific genius was required to invent the telescope shows that it was long overdue. The need for it had always existed, but nothing was done because it was not thought to be realizable. The means of making it had in fact been available for some 300 years. It seems, however, to have required the mere quantitative concentration of optical manufacture that went with the greater wealth of the sixteenth century to bring about its discovery by chance.

GALILEO GALILEI

The telescope was to prove the greatest scientific instrument of the age. The bare news of it reaching the ears of the professor of physics and military engineering at Padua, Galileo Galilei (1564–1642), determined

122. Portrait of Galileo Galilei (1564–1642) from the frontispiece of his *Istoria* of 1613 and reproduced again in his *Il Saggiatore* of 1623. The cherub at the top left holds Galileo's geometrical and military compass, and the one on the right holds what may have been his first design of a telescope.

him to make one himself and turn it on the heavens. Galileo was already a convinced Copernican, as well as being deeply interested in the movements of pendulums and the related problems of the fall of bodies. In the first few nights of observation of the heavens he saw enough to shatter the whole of the Aristotelian picture of that serene element. For the moon, instead of being a perfect sphere, was found to be covered with seas and mountains; the planet Venus showed phases like the moon; while the planet Saturn seemed to be divided into three. Most important of all, he observed that around Jupiter there circled three stars or moons, a small-scale model of the Copernican system, which anyone who looked through a telescope could see for himself.

With his keen sense of publicity and of the material value of his discoveries, which he found in no way incompatible with the pure joy of discovery, Galileo immediately tried to sell the titles of these stars in succession to the Duke of Florence (a Medici), to the King of France, and to the Pope, but the celestial honours seemed too expensive to all of them. Later, when the more practical end of using their motion to determine longitude at sea occurred to him, he tried to sell the secret to the King of Spain and the States General of Holland, who had both offered prizes for the discovery of the longitude, but still found no takers.[4.33.187]

These attempts, however, were to Galileo mere side-shows. He sensed at once the really revolutionary character of the new observations. Here he had for everyone to see the very model of Copernicus' system in the sky. This was knowledge not to keep but to broadcast. Within a month, in 1610, he had published what was clearly a scientific best seller, *Siderius Nuntius*, i.e. 'Messenger from the Stars', in which his observations were set out briefly and plainly. It created a great sensation and still it met with no immediately unfavourable reaction. The trial was not to be for another twenty-four years, and while a qualified condemnation of Copernican views was pronounced in 1618, it placed no obstacles to their being considered as a mathematical representation of the motions of the heavens. A few hard-bitten Aristotelians refused to look through the telescope, as they knew perfectly well what was in the heavens by the exercise of sheer reason. As long as reason and observation could be kept in different spheres of discourse there would be no trouble.

THE FALL OF BODIES: DYNAMICS

But Galileo felt it was not sufficient to have verified by observation the aesthetic preference of Copernicus. It was also necessary to justify it by explaining how such a system could exist, and by removing the objections which both philosophy and good sense had raised to it in the past.

It was necessary to explain how the rotation of the earth could occur without a mighty wind blowing in the opposite direction and how bodies projected through the air would not be left behind. This meant a serious study of bodies in free motion, a problem which had already become of great practical importance in relation to the aiming of projectiles.

By that time the impetus theory of Philoponos (p. 258), passed on by the Arabs and elaborated by the Parisian Nominalists (pp. 300 f.), was gaining acceptance. The projectile, on leaving the gun, was supposed to be endowed with an impetus or *vis visa* which for a while destroyed its natural propensity to fall downwards. Tartaglia (1500–1557), Benedetti

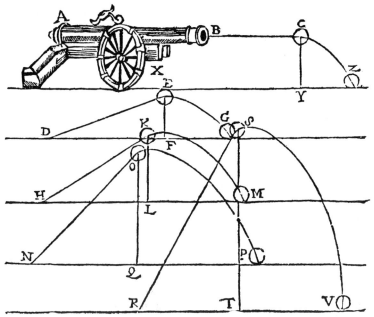

123. Only in the Renaissance was the Aristotelian doctrine of the path of a projectile overthrown. It was discovered that it moved in a parabola rather than in two straight lines. This woodcut from *Instrumentos Nuevos de Giometria* by Cespedes, published in 1606, shows the situation before Galileo's work was fully known and developed but when there was dissatisfaction with Aristotelian ideas. The intermediate theoretical stage when the theory of impetus was widely held gave rise to the straight line portion at the beginning of the trajectory shown here.

(1530–90), and others in the sixteenth century had elaborated this explanation by inserting between the violent rise of the projectile and its natural fall a circular mixed motion producing a trajectory that for the mortar bombs of the period was not too bad an approximation. What it lacked, however, was any logical or mathematical justification.[4.49; 4.101]

EXPERIMENTAL PHYSICS

Galileo succeeded, where others had failed, in formulating a mathematical description of the motion of bodies. This was to be the major work of his life, expressed fully only in his *Dialogues on Two New Sciences*, published after his condemnation but implicit in the *Dialogue Concerning the Two Chief Systems of the World*, which was to be the immediate cause of his conflict with the Church. Galileo proceeded to question all the accepted views and to do so by the new method, the method of experiment. Whether or not in fact he dropped weights from the top of the tower of Pisa is not the essential point; we know that he used both the pendulum and the inclined plane to make accurate measurements of the fall of bodies.

These were almost but not quite the first experiments of the modern science. They differed from the experiments of the thirteenth-century scholars mainly in being exploratory rather than illustrative, and even more by their quantitative character, which could be fitted into mathematical theory. Galileo himself showed a transitional attitude towards his own experiments. He once stated that he carried them out not to convince himself but to convince others. He was superbly confident in his power to interpret Nature by reason. In this sense they were rather demonstrations than experiments. Nevertheless he really did carry them out, unlike the ideal, paper experiments which befog modern physics; and what is more, when they gave results he did not expect, he did not reject them but turned back to question his own arguments, thereby showing the essential humility before fact that is the hall-mark of experimental science.

The mathematical interpretation of Galileo's experiments on falling bodies here proved to be far more difficult than the experiments themselves. The idea that had to be grasped was that a body that was changing its speed all the time could have any particular speed at a given moment. As a matter of fact Galileo went wrong to start with and assumed that speed was gained in proportion to the distance gone through by the body, whereas, as he himself was later to conclude, it depended directly on the *time* the body was falling.[4.101] To understand the falling of bodies, and consequently both the motions of cannon balls in the air and of the moon in the sky, it was essential to grasp the very difficult physical idea of the velocity at a moment of time. This corresponds to the mathematical idea of a differential, dx/dt: the ratio of two quantities that remains constant even if the quantities themselves become vanishingly small. Galileo used these ideas without precisely formulating them. By combining exact experiment and mathematical analysis, he solved the relatively simple problem of the fall of bodies, showing that in the absence of air they would follow a parabolic path. In doing so he provided the first clear example of the methods of modern physics, which were to have such an extraordinarily successful development in succeeding centuries. Indeed, the exact physical method he initiated has, until very recently, been taken as *the* basic method of science, one to which all other science may in the end be reduced.

THE RENAISSANCE OF MATHEMATICS

The achievement of Galileo and Kepler was possible because they were masters of the new *mathematics* that had blossomed with the Renaissance. Vieta (1540–1603) had taken the decisive step of making all algebraic argument *symbolic* by using letters for both known and unknown quantities not only in algebra but also in trigonometry. This purely technical device enormously speeded up calculations and removed the confusion that words inevitably induce. Thanks to his work, as well as that of Cardan (1501–76) and Tartaglia, algebraic methods could be employed for dealing with any problem where quantities could be reduced to numbers. The old Greek geometry still retained its prestige, especially since the recovery of the works of Archimedes, first edited by Tartaglia in 1543; but numerical calculations could be dealt with far more easily by the methods of algebra. An enormous practical step forward was registered when Simon Stevin (1548–1620) introduced decimals in 1585, and Napier (1550–1617) logarithms in 1614. By shortening computations by a large factor it effectively multiplied the number of working astronomers and physicists.

To complete the chain of argument it was necessary for Galileo to link mathematics with mechanics. How to do this was a major preoccupation throughout his whole scientific life. Leonardo was groping after a quantitative approach to mechanics; Galileo, with the advantages of better experiments and a more applicable mathematics, fully grasped it. He became one of the founders of scientific engineering. Another was the same Simon Stevin of Bruges, the first great engineer of the new Holland,

124. Diagram for his analysis of the strength of a beam, from Galileo's *Discorsi e Dimonstrazioni Matematiche intorno à due nuove scienze*, Leyden, 1638.

who took a large part in the war of liberation. He was responsible for the laws of composition of forces and for the foundations of quantitative hydraulics.

STATICS AND DYNAMICS: PRIMARY AND SECONDARY QUALITIES

A full understanding of the movement of massive bodies requires a treatment of forces first in equilibrium, as in *statics*, then out of equilibrium, as in *dynamics*. These were the 'Two New Sciences'[4.70] in which Galileo laid the foundations not only of the laws of motion but also of

125. The frontispiece of Galileo's *Dialogo sopra i due massimi Sistemi del Mondo* of 1632. The book, which supported the heliocentric theory of Copernicus, was condemned by the Roman Catholic Church. On the left is Aristotle, in the centre Ptolemy, and Copernicus is on the right.

the mathematical theory of the strength of materials, which he based on discussions with master shipwrights.

Galileo stated more clearly than anyone before him that the necessary and intrinsic properties of matter – the only ones in fact that could be dealt with mathematically, and therefore with any certainty – were extension, position, and density. All others, 'tastes, smells, colours, in regard to the objects in which they appear to reside are nothing more than mere names. They exist only in the sensitive body. . . .' This was not understood by the advocates of the new science as a limitation, but as a programme of reduction of all experiments to the primary qualities of 'size, shape, quantity, and motion'.

THE DESTRUCTION OF ANCIENT COSMOLOGY

To win general recognition for his new mathematical–mechanical science Galileo had first to destroy the Ptolemaic system of the heavenly spheres and with it, as he himself saw clearly, the whole Aristotelian philosophy which for nearly 2,000 years had been the foundation not only of the natural but also of the social sciences. He was particularly suited for the task, as he had seen Aristotelian philosophy at its best in Padua. He was no outsider, but was able to refute the master by his own logic in a way which scholars could not ignore, however much they might disapprove. Implicitly all his work was a protest against Aristotelians, but his first explicit blast came in 1632 in his polemical book *Dialogue Concerning the Two Chief Systems of the World, the Ptolemaic and the Copernican*, which he dedicated to the Pope. Here, not in learned Latin but in Italian for all to read, he mercilessly criticized and ridiculed the officially held views on the most important of subjects. This was the first great manifesto of the new science.[4.69]

THE TRIAL OF GALILEO

The challenge he put down could not be ignored and led directly to the famous trial. Galileo had made as many enemies in science as in the Church, and with the publication of the *Dialogue* they redoubled their denunciations of him. It is difficult to realize now why such an academic point as the motions of the earth and the planets should have caused such a violent struggle, but in those days far more was seen to be at stake. After centuries of violent disputes, and at the cost of the greatest intellectual effort, the Christian–Aristotelian compromise had been hammered out. Even the doctrinal quarrels of the Reformation had not shaken it. If the challenge in one essential aspect, the constitution of the heavens, was ignored, how much further might not the attack be pressed ?

Already ardent Copernicans, such as Bruno and Campanella (1568–1639), had drawn conclusions from the new knowledge that threatened the stability of the Church, government, public morals, and property itself (pp. 308 f.). Bruno had been burnt, Campanella was imprisoned for years; but with Galileo it was a different matter: he had scientific prestige and powerful friends, his Catholicism was not in doubt, and, except in science, he was not a revolutionary.[4.38; 4.57]

The trial was necessarily carried on in terms of the ideas and mode of reasoning of the Church and not those of Galileo, so that the result was a foregone conclusion. But the interesting fact is that the proceedings of the trial were kept secret, most probably because of the danger that their publication would reveal not the severity of the judges but their comparative leniency.[4.151] The Pope and the Curia were more anxious about the possible reactions of the fanatical diehards of the Church than about those of the scientists. Galileo was condemned and forced to make his famous recantation, but he suffered a merely nominal imprisonment in the palace of one of his friends. In his retirement he was able to complete his work on dynamics and statics and to publish it in the latter years of his life.

Nevertheless, the event of the trial marked an epoch, for it dramatized the conflict between science and religious dogma. Through its effective failure, for the verdict was badly received by nearly all the learned, even in Catholic countries, it gave enormous prestige to the new revolutionary experimental science, especially in countries that had already overthrown the authority of Rome. Galileo's achievement appears as the culmination of the attack on the old cosmology. From then on it was quietly dropped, and practical astronomers used the Copernican–Keplerian model of the solar system. Forty years later the observational laws of Kepler were to be combined with the dynamics of Galileo in Newton's theory of universal gravitation.

MAGNETISM: NORMAN AND GILBERT

One further physical clue which led to that synthesis was the experimental study of magnetism, known to the world through the publication in 1600 of *De Magnete* by William Gilbert, Queen Elizabeth's physician. The experimental discovery on which it was based, that of the dip of a balanced needle, had already been noticed by Hartmann (1489–1564) in 1544 and studied in detail by Robert Norman (*fl.* 1590), a mariner and compass maker, and one of the first scientists with neither gentle birth nor book-learning. He is fully conscious of his rights as he sets them out in the preface of his *The Newe Attractive* (1581):

... yet I meane God-willing, without derogating from them, or exalting my-self, to set down a later experimental truth found in this stone, contrarie to the opinions of all them that have heretofore written thereof. Wherein I mean not to use barely, tedious Conjectures or imaginations: but briefly as I may to passe it over, grounding my Arguements only upon experience, reason and demonstration, which are the grounds of Art. And albeit, it may be said by the learned in the Mathematicalles, as hath beene already written by some, that this is no question or Matter for a Mechanitian or Mariner to meddle with, no more than is the finding of the longitude, for that must bee handled exquisitely by Geometricall demonstration, and Arithmetical Calculation: in which Artes, they would have all Mechanitians and Sea-men to be ignorant, or at leaste insufficientlie furnished to performe such a matter, alledging against the latin Proverb of Apelles, *Ne sutor ultra crepidam*. But I doe verily thinke, that notwithstanding the learned in those Sciences, being in their studies amongst their bookes, can imagine great matters, and set downe their farre fetcht conceits, in faire showe, and with plawsible words wishing that all Mechanitians were such, as for want of utterance, should be forced to deliver unto them their knowledge and conceites, that they might flourish upon them, and applye them at their pleasures: yet there are in this land divers Mechanitians, that in their severall faculties and professions, have the use of those Artes at their fingers endes, and can apply them to their severall purposes, as effectually and more readily, than those that would most condemne them.

I have quoted this at length as a manifesto of the challenge of new craftsmen to the old scholars. It finds an echo in the polemics of Gabriel Harvey (1545–1630), the rope-maker's son, the friend of Spenser who claimed the same rights in literature that were shortly to be vindicated by the glover's son William Shakespeare. Harvey writes: [4.84]

He that remembreth Humfrey Cole, a Mathematicall Mechanician, Matthew Baker a ship-wright, John Shute an Architect, Robert Norman a Navigator, William Bourne a Gunner, John Hester a Chimist, or any like cunning, and subtile Emperique, (Cole, Baker, Shute, Norman, Bourne, Hester, will be remembered, when greater Clarks shal be forgotten) is a prowd man, if he contemne expert artisans, or any sensible industrious Practitioner, howsoever unlectured in Schooles, or unlettered in Bookes ... and what profounde Mathematician, like Digges, Hariot, or Dee, esteemeth not the pregnant Mechanician? Let every man in his degree enjoy his due: and let the brave engineer, fine Daedalist, skilful Neptunist, marvelous Vulcanist, and every Mercurial occupationer, that is every Master of his craft, and every Doctour of his mystery, be respected according to the uttermost extent of his publique service, or private industry.

Nevertheless, the scholars still had important tasks to perform. They had to transmit the knowledge of the past to the new craftsmen–scientists till these could all learn to stand on their own feet, and they had, through their connexions with rank and wealth, to assure recognition and support for the new sciences. Gilbert fulfilled both functions admirably. His *De Magnete*, though full of strong invective in Latin as any Norman or Harvey could use in English against the blindness of the old philosophers, was also so well buttressed with scholarship as to compel assent of the whole learned world, though Norman's book must have been of more use to sailors and compass makers.

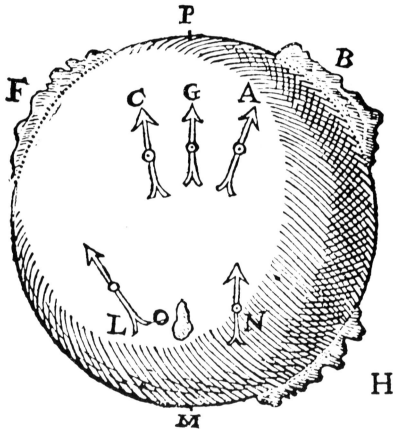

126. Gilbert's idea of the Earth as a magnet. An experiment using a terrala or globe-shaped magnet with lumps of iron to represent mountains, and showing the north-seeking property of the magnetic needle. From William Gilbert, *De Magnete*, London, 1600.

De Magnete is a great book in itself, and as an exposition of the new scientific attitude. Gilbert did not confine himself to experiments: he drew from them new general ideas. The one that struck most at the imagination of his time was his idea that it was the magnetic virtue of *attraction* that held the planets in their courses. It provided the first physically plausible and completely non-mythical explanation of the ordering of the heavens. It certainly made it easier for Newton in his argument against the physically minded scientists who could conceive force only by the impulsion of material bodies in contact.

THE MECHANICS OF THE HUMAN BODY

It was not, however, only in the skies and in the stones that the old views were yielding to the new. At the same time an equally successful attack was being made on the inner universe – the nature of man's body. The Aristotelian world-picture was essentially centred on the earth and man. Man at the centre of the universe was supposed to be in direct contact with all its parts by means of influences and spirits that connected him with the planetary spheres. He was a little world in himself – a microcosm. Its detailed workings had been elaborated by the Greek doctors ending in Galen, whose description of the organs of the human body had become as canonical as Ptolemy's description of the heavens. The new anatomy of the Renaissance, particularly the work of Vesalius, showed that Galen's picture must be wrong; but the alternative explanation could be found only by a totally new approach to the problem, one which combined anatomy with the new Renaissance interest in machinery – bellows, pumps, and valves – and could derive from them a new experimental physiology.

HARVEY AND THE CIRCULATION OF THE BLOOD

This was to be the work of William Harvey (1578–1657), an Englishman of good family, trained in Padua, and so able to combine the Italian anatomical tradition with the new interest in experimental science that was beginning to find its way into England.[4.68] What Harvey sought was the mechanical explanation of the movements of the blood in the body. His *Exercitatio Anatomica de Motu Cordis et Sanguinis in Animalibus* published in 1628 is the record of a new kind of anatomy and physiology. No longer is it mere dissection and description, but an active investigation, a piece of hydraulic engineering research carried out by means of practical experiments. Harvey had a difficult case to prove; he had to overcome the disability of being, as it were, a Copernicus, forced to deduce his new system without a Galileo to confirm it by visible evidence. He could prove logically that a circulation must exist, because blood went out of one side of the heart and came back into the other – far more blood than could be currently accommodated in the body itself. But he could not *see* how it got from one side to the other. The fine capillary (hair-like) vessels through which it flowed were to be demonstrated later by Malpighi (1628–94), using the other new optic glass, the *microscope*.

What Harvey established by his close reasoning from experiment had the same revolutionary effect on ancient and Galenic physiology as the discoveries of Galileo and Kepler had on Platonic and Aristotelian astronomy. He showed that the body could be looked at as a hydraulic machine and that the mysterious spirits which were deemed to inhabit it had no place to live in (p. 223). His own views, however, remained more Copernican and Keplerian than Galilean, with a strong sense of the parallelism of the body to the world.[4.7] He writes, for example:

> So the heart is the beginning of life, the Sun of the Microcosm, as proportionably the Sun deserves to be call'd the heart of the world, by whose vertue and pulsation, the blood is mov'd, perfected, made vegetable, and is defended from corruption and mattering; and this familiar household-god doth his duty to the whole body, by nourishing, cherishing, and vegetating, being the foundation of life and author of all.[4.8.56]

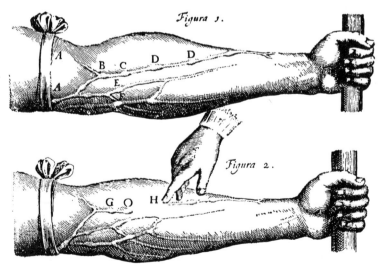

127. An engraving to illustrate the valves in the circulation of the blood. From William Harvey, *De Motu Cordis*, Frankfurt, 1628.

Thus he puts the heart in the body in the same royal, central place as the sun in the universe. Harvey's beautiful demonstration of the mechanics of circulation gave great weight to the idea that an organism was a machine, though it turned out later that it was one far more complicated than the men of the sixteenth and seventeenth centuries imagined.

Harvey's discovery had, however, very little immediate effect in medicine, apart from justifying methods to prevent people from bleeding to death, already practised by field surgeons like Paré. It was, however, absolutely necessary as a foundation for any rational physiology. The picture of the organism, as it grew up from Harvey's work, is that of a set of organs, what might be called 'irrigated fields', provided with a circulation which keeps every part in communion with the rest in a nutritive and chemical way.

CHEMISTRY

This understanding was to be delayed, for the chemical advances of the century from 1540 to 1640 had not been startling. The only man of first-class mind to have occupied himself with it was van Helmont (1577–1644), a nobleman trained in medicine and a follower of Paracelsus, whose mystical views he approved though he had no need for his bombast. His chemical ideas hark right back to the Ionians, believing the only elements to be air and water. But this was not so much a philosophic hypothesis as an experimental conclusion, for he grew a willow tree from a seed in a pot to which water only had been added. He was also the first to name and study gas – or chaos – pointing the way to the triumphs of later chemistry. For the rest, chemistry pursued its slow and steady course of widening its basis of experience, improving the accuracy of its measurements, and increasing the scale of its operations, particularly in the distillation of spirits.

7.6 The New Philosophy

The two great and hard-won discoveries of the rotation of the planets and of the circulation of the blood were both securely established by 1642, the year Galileo died and Newton was born. The first intellectual object of the scientific revolution had been achieved: the classical world-picture had been destroyed, though only the bare outlines of a new one had been put in its place. In doing so new means for understanding and conquering Nature had been found, but little had as yet emerged that could be claimed to be of general practical use. The telescope itself was a technical rather than a scientific invention. Before the effects of the revolution in thought could make themselves felt in practice, it was necessary that the possibilities the new science offered should be brought home not only to the learned but to the new class of enterprising people who were making their own political revolution – merchants, navigators, manufacturers, statesmen, and the early and progressive capitalists. Galileo had started to do this, but he was living in a country that had already lost its *élan* and that was rapidly being frozen into reaction by the Counter-Reformation.

128. Rene Descartes (1596–1650): an engraving of the portrait by Frans Hals the Elder, now in the Louvre.

o

THE PROPHETS: BACON AND DESCARTES

Two men from the less cultured but far more active northern countries were to take on the task – Bacon and Descartes. These two major figures stood at the turning point between medieval and modern science. Both were essentially prophets and publicists, men who had seen a vision of the possibility of knowledge and were making it their business to show it to the world. Both were universal in scope, though their approaches to knowledge were very different. Temperamentally, too, it would be difficult to find two more different people than the shrewd, self-seeking, and afterwards rather pompous lawyer, always at the centre of public affairs, and the intensely introspective, solitary ex-soldier of fortune. Each too is characteristic of the nature of the scientific revolution in his own country.

Bacon emphasized the essentially practical side of the new movement, its applications to the improvement of the arts, its usefulness in bringing about a more common-sense appreciation of the world around them. Living as he did in the courts of Elizabethan and Jacobean England, he found that his difficulties arose not so much from the existence of rigid systems of thought as from the need to lay solid institutional foundations for a new and generally acceptable philosophy. This was put forward not only to replace the older views, but also to put in order the chaos of speculations that the Reformation in England had produced. Descartes, on the other hand, had to fight against a medieval system of thought entrenched in the official universities of France, and only succeeded by using a logic that was clearer and intellectually more compelling than theirs.

THE 'NOVUM ORGANUM' AND THE 'DISCOURS DE LA MÉTHODE'

Both thinkers were preoccupied with methods, though their ideas of scientific method were very different. Bacon's was that of collecting materials, carrying out experiments on a large scale, and finding the results from a sheer mass of evidence – an essentially *inductive* method. Descartes, on the other hand, believed in the rapier thrust of pure intuition. He held that with clarity of thought it should be possible to discover everything rationally knowable, experiment coming in essentially as an auxiliary to *deductive* thought. The major difference, however, was that while Descartes used his science to construct a *system* of the world, a system which, though now almost forgotten, was able in its time completely to supersede that of the medieval schoolmen, Bacon put forward no system of his own but was content to propose an *organization* to act as a collective builder of new systems. His function as he saw it was only to provide the builders with the new tool – the logic of the *Novum Organum* – with which to do it.

In this sense they were strictly complementary. Bacon's concept of organization led directly to the formation of the first effective scientific society, the Royal Society. Descartes' system, by breaking definitely with the past, put up a set of concepts which could be the basis of argument about the material world in a strictly quantitative and geometric manner.

The thoughts of both philosophers were, nevertheless, inevitably deeply tinged with medieval ideas, though each in a different way. Francis Bacon belonged to the tradition of the encyclopedists, of his namesake Roger Bacon and of Vincent of Beauvais (p. 306), or, tracing farther back, of Pliny and Aristotle himself. He was, first and foremost, a naturalist in interest and had no knowledge of or sympathy with the new mathematical philosophy. His method was largely negative, based on the avoidance of the 'idols' or false lures of ideas that had led the old philosophers astray. His imaginary House of Solomon in his *New Atlantis* [4.30] was a kind of universal laboratory, an idealization of the actual observatory of Tycho Brahe at Uraniborg. It was in turn to be the inspiration of later scientific institutes. Though a believer in experiments, Bacon was not an experimenter himself, and he never fully understood the process of abstraction and reduction needed to extract truth from complex situations which Galileo was already using so magnificently. He thought that systematic, common experience, purged of the pernicious ideas of the Ancients, would suffice for knowledge. His scientific beliefs were not original but drawn from reading, particularly of Telesius, whom he criticized but called 'the first of the moderns'.

Telesius (1509–88), an Italian scholar, was the first to break absolutely with Aristotle by setting up a rival system. His great contribution was the abandoning of Aristotle's formal and final causes and the retention of only material and efficient causes (pp. 201 f.). In this he was followed by all later science. His own views recall those of Anaximenes. His universe worked by means of the inner powers of heat and cold. This was an anticipation of the doctrine of energy and included some idea of conservation, but was not much more advanced quantitatively than the Yang and the Yin of Chinese philosophy (p. 176).

From the very beginning of his career Bacon set out to preach the doctrine that 'The true and lawful end of the sciences is that human life be enriched by new discoveries and powers.' He saw himself not so much as a scientist and inventor, but as an inspirer of science and invention: 'I have only taken upon me to ring a bell to call other wits together.' In his admirable study of Francis Bacon, Professor Farrington quotes: [4.65]

Nowe among all the benefits that could be conferred upon mankind, I found none so great as the discovery of new arts, endowments, and commodities for the bettering of man's life. For I saw that among the rude people in the primitive times that authors of inventions and discoveries were consecrated and numbered among the gods. And it was plain that the good effects wrought by founders of cities, law-givers, fathers of the people, extirpers of tyrants, and heroes of that class, extend but over narrow spaces and last but for short times; whereas the work of the Inventor, though a thing of less pomp and show, is felt everywhere and lasts for ever.

But above all, if a man could succeed, not in striking out some particular invention, however useful, but in kindling a light in Nature – a light which should in its very rising touch and illuminate all the border-regions that confine upon the circle of our present knowledge; and so spreading further and further should presently disclose and bring into sight all that is most hidden and secret in the world – that man (I thought) would be the benefactor indeed of the human race – the propagator of man's empire over the universe, the champion of liberty, the conqueror and subduer of necessities.

Bacon was taken to be, and rightly, the first great man who had given a new direction to science and who had linked it definitely once more to the progress of material industry.[4.54]

With his empirical bent Bacon was inevitably an opponent of all pre-determined systems in Nature; he believed that, given a well-organized and well-equipped body of research workers, the weight of facts would ultimately lead to truth. Descartes' method, on the other hand, was a more direct successor of that of the schools, with this absolute difference: that it was not *their* system that he wanted to establish but *his own*. In this he exhibited that individual arrogance which was one of the great liberating features of the Renaissance, the same arrogance that expressed itself in the great navigators, in the *conquistadores*, in all the defiances of authority that characterized the end of the feudal period and the beginning of one of individual enterprise.[4.6]

Unconsciously, Descartes' system incorporated very much of the system which he wished to destroy. There was the same insistence on deductive logic and self-evident propositions, but starting with these he used the *mathematics*, of which he was a master, to arrive at conclusions far beyond the reach of his medieval or even of his classical predecessors. His major mathematical contribution was the use of co-ordinate geometry, by which a curve could be completely represented by an equation relating the values of the co-ordinates of its points referred to fixed axes. This was more than the mapping of geometry. It broke down the old distinction between the Greek science of the continuum – *geometry* – and the Babylonian–Indian–Arabic calculus of numbers – *algebra*.

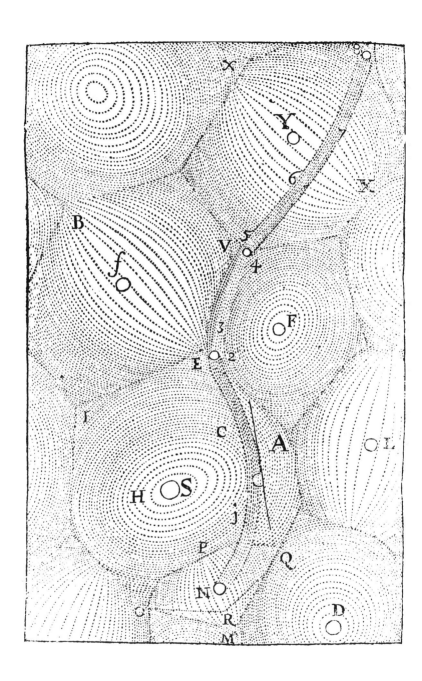

129. Descartes' universe was a plenum of material particles arranged in vortices. The Sun, he believed, was the centre of one such vortex, with the planets orbiting round it. From René Descartes, *Epistolae*, Elzevir, Amsterdam, 1668.

Henceforth their two powers would be joined to attack problems never before attempted.

In his attack on the old philosophy Descartes was as canny as he was courageous. He had no desire to enter into a head-on conflict with organized religion, a conflict that had led to the condemnation and burning of Bruno in Catholic Rome and that of Servetus in Calvinist Geneva. He was prepared to be accommodating, and he hit on an ingenious method of doing so which was to make science possible for several centuries at a cost which we are only now beginning to feel.

PRIMARY AND SECONDARY QUALITIES

Descartes formulated, more precisely than anyone before him, the division of the universe as we see it, into a part which is physical and one which is moral. Other philosophers, going back to the Arabs and the Scotists in the Middle Ages including Roger Bacon, as well as Francis Bacon himself, had made reservation of the knowledge that only came by faith or revelation (pp. 300 f.), but this pious reservation was *ad hoc* and was open to the objection that it implied that God was irrational. With Descartes this separation became an integral and rational part of philosophy. It was a logical consequence of his reduction of sensory experience first to mechanics and then to geometry. As with Galileo, extension and movement were the only physical realities that he recognized as 'primary'; other aspects of existence, such as colours, tastes, smells, were referred to as 'secondary' qualities. Beyond these stretched a region even more unapproachable by physics, the region of the passions, of will, love, and faith. Science, according to Descartes, concerned itself mainly with the first set – the measurables, the bases of physics; and to a lesser extent with the second; but not at all with the third, as they lay in the realm of revelation.[4.58] To Descartes, animals, including men, were in themselves merely machines. Obviously there must be some connexion between the purely mechanical man, operating his limbs according to physical principles, and the rational spirit and will dwelling within him. Descartes made the naïve, but apparently quite serious, suggestion that the connexion could be through the small gland at the top of the skull – the pineal body – the relic of a pair of eyes in our reptilian ancestors, but having now no apparent function and therefore quite reasonably, if not the seat, at least the point of entry of the rational soul.

THE SEPARATION OF RELIGION AND SCIENCE

The effect of Descartes' division ever since was to enable scientists to carry on their work free from religious interferences so long as they did not trespass into the religious sphere. This was, of course, very difficult to avoid or refrain from, but nevertheless, it had the effect of producing the type of pure scientist who kept out of fields where he was likely to be involved in controversies of a religious or political kind. To a certain extent Descartes himself must have done this, because the story goes that when he had ready his *System of the World* he heard the news of the trial of Galileo and realized that it simply would not do as it was. The Church was clearly determined that the Aristotelian–Thomist system was necessary to secure the truths of the Faith and was not going to tolerate any other system that might put them in question. Descartes consequently set himself to the task of showing that his systems could prove the existence of God quite as well as, if not better than, the older philosophies. From his famous first deduction *Je pense donc je suis* – 'I think, therefore I am' – he drew the conclusion that as all men can conceive something more perfect than themselves, a perfect being must exist. Descartes' system was so carefully guarded against theological attack that, in spite of protests from the universities, it was accepted in that most Catholic country, France, within his own lifetime and for a century after his death.

Descartes' system was, however, in spite of its wealth of mathematical and observational content, essentially a magnificent poem or myth of what the new science might be. That was at the same time its attraction and its danger. It was a mixture of conclusions soundly based on experiment with those deduced from first principles chosen, according to Descartes' celebrated *Method*, only on account of their *clarity*. The pursuit of that clarity has been the ornament and the limitation of French science ever since. Where in the state of knowledge it was admissible, as in eighteenth-century dynamics and chemistry and in nineteenth-century bacteriology, it could be used to put in order whole fields of genuine but chaotic knowledge. Elsewhere it tended to degenerate into arid commonplaces and false simplifications.

To a certain extent Descartes himself recognized the limitation of a one-man enterprise in philosophy, and realized that the proper establishment of the system of the world would require the co-operation of many minds. In the *Discours de la Méthode*, he says, speaking of experiments:

I see also that there are so many of them that neither my hands nor my wealth, even if I had a thousandfold what I have, would serve me for this end. . . . What I had to show by my treatise was the utility that the public can gain from it and that I should oblige all those who desire the good of mankind, that is to say all those who are really virtuous and not only pretend to be, to communicate to me their own results and to help me in the researches that still have to be made.

In another place he says, to justify the publication of his own conclusions:

They showed me that it is possible to arrive at knowledge very useful to life: and that instead of this speculative philosophy that is taught in the schools one can find a practical philosophy by which *knowing the force and action of fire, water, air, the stars, the heavens, and all other bodies that surround us as distinctly as we know the different trades of our craftsmen, we could employ them in the same way to all uses for which they are appropriate and thus become the masters and possessors of Nature.* This is not only desirable for the invention of an infinity of artifices which would enable us to enjoy without trouble all the fruits of the earth – but principally in the preservation of Health.

Thus in his ultimate objective Descartes did not differ much from Bacon, for whom he had, in any case, the greatest admiration. Bacon and Descartes between them raised the status of experimental science to an esteem in polite circles comparable to that of literature. From their time on the new natural philosophy, and not that of the schools, was the centre of interest and discussion. Indeed, after another 200 years or so it had just about fought its way into the universities of England.

The time was now ripe for a great expansion of this natural science and its first fruits. In the next period from 1650 to 1690 the 'Great Instauration' – or as we would say Reconstruction – of which Bacon dreamed was at last to take place.

I entrust men to believe that it is not an opinion to be held but a work to be done: and to be well assured that I am not labouring to lay the foundation of any sect or doctrine, but of human utility and power.

7.7 The Third Phase: Science Comes of Age 1650–90

The third and definitive phase in the establishment of modern science was reached in the latter half of the seventeenth century. Intellectually, as we have seen, the ground had been prepared for it by the overthrow of the feudal-classical theories in the previous hundred years. Though this made further advance and consolidation of science possible, it was not the only, nor the main, cause of the outburst of activity which, in less than fifty years, virtually created modern science in most of its fields. This intense growth was more concentrated than at any time before or since. The principal foci were London and Paris, for the active scientists of Italy and Holland found no such centres of expression in their own countries, while those of central and eastern Europe had not yet come into action.

The condition which made this rapid growth possible and favoured its concentration was first and foremost the establishment in Britain and France of stable governments in which the upper bourgeoisie had a dominating, or at least an important, part. In Britain the Civil War had brought about a real revolution, in which the richer merchants with the help of the townsmen and small landowners had won power from the king and the landed nobility. But these groups, after their triumph, soon quarrelled. The small men had a distressing tendency to democracy and economic equality,[4.170] and as soon as Cromwell was out of the way the merchant interest arranged a compromise with the landlords in which King Charles II came in as the first constitutional monarch. The merchants still dominated the economy, but a new class of manufacturers were making their first appearance, drawn partly from the ranks of the merchants, partly from those of the skilled craftsmen. The great increase of manufacture and trade that followed the end of the Civil War, together with the new possibilities of navigation, kept mechanical invention at a premium. The time and place were in every way most propitious for the growth of science.

Holland, though immensely rich, was by the middle of the century past her prime. Sixty years had passed since the revolution which had ended the rule of Spain. The popular support that had secured the independence of the country had been largely dissipated, and the government was in the hands of a combination of wealthy merchants and landlords. Soon, exhausted by commercial wars and without adequate manufactures, Holland was to prove too weak to maintain her leading place. Already by the end of the century some of the most able Dutchmen took service abroad, particularly in the development of Britain under William of Orange, while Holland's greatest scientist, Christian Huygens, did most of his work in Paris as a member of the French Academy.

In France, on the other hand, the Revolution was still in the future. The strength of feudalism and of the Church had been shown in the crushing of the Huguenots; but this was a slow process and was only fully effected by the revocation of the Edict of Nantes in 1685. Nor could this vigorous and expanding country, then by far the largest and richest in Europe, stand aside from the general economic development. A compromise was patched up by which the nobles bartered part of their power for tax exemption, pensions, and pageants at Versailles. The executive power was centred on the king, but his State machine was bourgeois throughout. It was largely run by intelligent lawyers, the *Noblesse de Robe*, from which many scientists were to come. Actually, the compromise only worked tolerably well in the early part of Louis XIV's personal reign (1661–83) under the direction of the business-like Colbert, and this coincided exactly with the great period of science.

The other countries of Europe played minor parts on the scientific stage: Germany and Austria had only begun to recover from the Thirty Years War (1618–48); the Inquisition neutralized Spain and Portugal almost completely; while in Italy the heirs of Galileo fought a gallant rearguard action against the forces of clericalism.[4.136; 4.151] Sweden, Poland, and Russia were still largely raw-material countries in the throes of a newly imposed serfdom and, though militarily strong, were only beginning to contribute to science at this stage.

LE GRAND SIÈCLE

After the great religious and political disturbances of the previous hundred years the latter half of the seventeenth century was a period of relative calm and active prosperity. Plagues and wars were constant but had surprisingly little effect on the work of the scientists. Nor did national rivalry seriously interfere as yet with their freedom of movement or communication. It was an age of conscious building of civilization – Le Grand Siècle – and the scientists were recognized and honoured as part of one common republic of letters. The governments and the ruling classes of all the leading countries had certain common interests in trade and navigation, and also in improvements in manufactures and agriculture. These interests were to furnish the motive power for the culminating achievements of this third phase of the Scientific Revolution, the first in which an organized and conscious effort was made to use science for practical ends.

This was the *fruit* which, thirty years before, Bacon had urged men to cultivate; and Bacon's methods, both those of experiment and of organization of research, were used to gather it. The men who were to do so were characteristic of their age and nations. In the place of the courtiers and university professors of the first two phases, depending for their livings on the favour of princes, the *virtuosi* of the mid seventeenth century were men of independent means, mostly merchants, middling landowners, and successful followers of the liberal professions – doctors, lawyers, and not a few parsons. They might seek royal patronage but they could count on little royal money for science; King Charles II never paid a penny to his Royal Society and never even managed to find time to visit it. The *virtuosi* had to finance science out of their own pockets. But these pockets were ample and were being rapidly filled by the great increase in trade, whose benefits now flowed into the very countries where science flourished. Some were even able to take on other scientists about the work. The Hon. Robert Boyle employed Hooke, a poor curate's son, just as Christian Huygens, lord of Zulichem in Holland, employed Denis Papin of Blois.

These men were competent and interested enough to carry out scientific research on their own; but as they became more numerous they tended to gravitate naturally together for discussion and interchange of knowledge, made all the easier by the commercial and levelling tendencies of the time. They went further: inspired by the propaganda of Bacon they began to think of a positive organization deliberately aimed at winning the secrets of Nature by a co-operative effort.

THE FOUNDATION OF SCIENTIFIC SOCIETIES

This third phase accordingly was the period of the formation of the first well-established scientific societies, the Royal Society of London and the French Royal Academy, which set themselves the task of concentrating on the central technical problems of the time, those of *pumping* and *hydraulics*, of *gunnery* and of *navigation*, while almost ostentatiously avoiding general philosophical discussions. It was particularly the navigational problems that furnished the stimulus to the advancement of science, because it was through the attack on these problems that the two elements of earlier science – mechanics and astronomy – were brought together in the great synthesis of Newton. In the latter part of this chapter I will try to trace some of the threads of experiment and argument that led to this synthesis. More important practical results were, however, to come from the study of the pump, which was to lead first to the discovery of the *vacuum*, then to that of the laws of gases – from which arose the steam-engine, as well as the pneumatic revolution of chemistry in the next century.

The establishment of science as a fully recognized factor in culture was definitive from the moment that scientific societies were formed. The idea of a scientific society was, as we have seen, a very old one. It found expression in the original Academy (pp. 196 f.), in the Lyceum (p. 198), and in the Museum of Alexandria (pp. 212 f.). Both Muslim and Christian universities were something of the same kind in their early stages, but by the seventeenth century it was evident that these could not fill the new needs. Something different was wanted and in due course appeared, partly in response to the inspiration of prophets of the new age like Francis Bacon, but even more as a formal recognition of spontaneous gatherings of men interested in science.

Among the prophets, John Amos Comenius (1592–1670), the last bishop of the Moravian Church, was an outstanding figure.[4.111] Approaching science as a part of universal education, to which he had devoted most of his life, he planned a 'Pansophic College' where the new experimental philosophy would be practised and taught. Driven out of Bohemia by the Thirty Years War he lived a wandering life, and was sought after by forward-looking governments because of his successful educational methods. It was beginning to be recognized by the statesmen of the new national States that an educated laity was needed to run the administration. In 1641 Comenius came, at the invitation of Parliament, to England, where he hoped to found his college. Though, owing to the difficulties of the time, he failed in this, his influence played some part in bringing the Royal Society into existence.[4.146]

Actually the earliest of the scientific societies were the Accademia dei Lincei at Rome (1600–30) and that of the Cimento at Florence (1651–67).[4.15] These, though they acted as models for societies elsewhere, arrived too late on the Italian scene to halt the factors inimical to science which soon led to their extinction. The Royal Society of London (1662) and the Académie Royale des Sciences in France (1666) were more fortunate. All arose originally from early informal gatherings of friends interested in the new sciences.

French scientists, among them Gassendi, who reintroduced the atomic theory, had been meeting at the house of a wealthy lawyer, Pieresc, at Aix-en-Provence as early as 1620.[4.41] The real centre of French science was, however, until his death in 1648, the cell of the minorite friar, Mersenne, himself no mean scientist. He was an indefatigable correspondent, acting as a kind of general post office for all scientists in Europe from Galileo to Hobbes.[4.106] Later, meetings were held in the house of another lawyer, Montmor, out of which the Royal Academy was ultimately to be formed.

Another promoter of a rather different type was Renaudot (*d.* 1679), a lively and combative doctor who, much to the horror of the faculty in Paris, set up a clinic giving free treatment to the poor. He combined this establishment with a lecture room for scientific meetings, a publishing house, and an employment agency, which largely paid for the whole

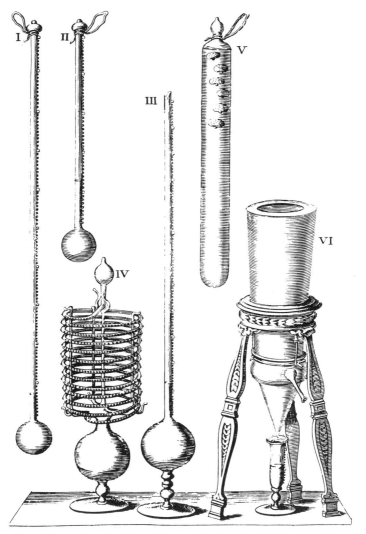

130. One of the earliest scientific societies was the Accademia del Cimento, founded at Florence in 1657. Two of Galileo's disciples, Vincenzio Viviani (1621–1703) and Evangelist Torricelli (1608–47), had laid the foundations some years before its formal institution. This engraving from *Saggi di naturali experienze fatte nell' Accademia del Cimento*, second edition, Florence, 1691, shows various kinds of thermometers (I–V) and a rain gauge (VI).

outfit. With the death of his protector Cardinal Mazarin in 1661, his enemies succeeded in shutting it down and putting an end to popular science in France for more than a hundred years.

In England the signal for the gathering of the new experimental scientists was the end of the Civil War in 1645. Most of them were Parliamentarian in sympathy, some Puritans, but had little to do with the actual fighting. The leading spirit of the group was John Wilkins, a clergyman of some adaptability in politics, marrying Cromwell's sister and ending as Bishop of Chester, but an unswerving supporter of the new philosophy. With him were associated the mathematician Dr Wallis, Dr Theodore Haak, a German refugee who was first to suggest the weekly meetings, and a number of doctors. After some preliminary meetings in London they settled down in Oxford in 1646. That loyal university had just been reformed by a Parliamentary Commission, and the empty chairs and headships of houses were filled with the new members of the 'Invisible College'. Until the Restoration in 1660 Oxford was to be, anomalously and somewhat unwillingly, the centre of the attack on Aristotle, who had been so revered there before and since. In Oxford the band was strengthened by the adherence of three young men of promise, the Hon. Robert Boyle, Sir William Petty, and Dr Christopher Wren, and also, though in a humbler capacity, Robert Hooke, the man who was to do most to make the Royal Society a success. As Thomas Sprat, one of the group, the future bishop of Rochester and historian of the society,[4.139; 4.140] wrote of those times:

Their first purpose was no more, then onely the satisfaction of breathing a freer air, and of conversing in quiet one with another, without being ingag'd in the passions, and madness of that dismal Age. And from the Institution of that *Assembly*, it had been enough, if no other advantage had come, but this: That by this means there was a race of young Men provided, against the next Age, whose minds receiving from them, their first Impressions of *sober* and *generous knowledge*, were invincibly arm'd against all the inchantments of *Enthusiasm* . . .

For such a candid and unpassionate company, as that was, and for such a gloomy season, what could have been a fitter Subject to pitch upon, then *Natural Philosophy?*

. . . *that* never separates us into mortal Factions; that gives us room to differ, without animosity; and permits us, to raise contrary imaginations upon it, without any danger of a *Civil War*.

Their *meetings* were as frequent, as their affairs permitted: their proceedings rather by action, then discourse; chiefly attending some particular Trials, in *Chymistry*, or *Mechanicks:* they had no Rules nor Method fix'd; their intention was more, to communicate to each other, their discoveries, which they could make in so narrow a compass, than an united, constant, or regular inquisition.

At first these amateur scientists merely met, discussed, showed each other experiments, and wrote letters to their absent friends or to their colleagues in other countries. The business of scientific communication and publication originated in these at first purely informal and then more regular letter-writings. Later, the need for a definite establishment was felt by the scientists both in England and France because, as they continued their work, they realized that it was likely to have considerable practical importance, and to carry it out they would have to have more money or more recognition.

The procedures differed according to the character of the economies of the two countries. In France, with its rigidly centralized government, it was natural that the establishment should be not only royally instituted but also royally paid. Colbert was setting up national industries in France, and it was accordingly not difficult to persuade him to found the Academy of Science to balance Mazarin's Academies of Literature and Fine Art. But then, ornament and show, just as much as commerce, were necessary to the glory of the kingdom of *le Roi Soleil*. The industries favoured by Colbert were those of silk-weaving at Lyons, pottery at Sèvres, and the Gobelin tapestries in Paris, all considered of importance comparable to the shipbuilding for the French Navy.[4.14]

In Restoration England, on the other hand, with its relics of republican independence and where the real wealth of the country was in the hands of the landed aristocracy and the merchants, royal patronage was all that was required. The Fellows of the new Royal Society paid for their own scientific investigations. The charge was one shilling per member per week. It was extremely difficult to collect and was hardly sufficient to pay the secretary and the curator, who 'shall be well skilled in Philosophical, and Mathematical Learning, well vers'd in Observations, Inquiries, and Experiments of Nature and art', and was obliged to 'furnish the Society every day they meete, with three or four considerable experiments, expecting no recompense till the Society gett a stock enabling them to give it'.[4.19; 4.26; 4.53; 4.83; 4.124]

The necessary consequence of official recognition of the societies was general conformity of ideas and avoidance of controversial issues in politics and religion. In France the Church grudgingly withdrew its insistence on Aristotelianism and accepted the compromise proposed by Descartes (p. 445). In Britain the same division of fields of interests came about in a different way. It arose from the troubles of the Great Rebellion in the middle of the seventeenth century and the desire of the early scientists to avoid the endless theological–political disputes that occupied most intellectuals in those times. In the draft preamble to the Statutes of the Royal Society written by Hooke in 1663, it was laid down that:

The business of the Royal Society is: To improve the knowledge of naturall things, and all useful Arts, Manufactures, Mechanick practices, Engynes and Inventions by Experiment – (not meddling with Divinity, Metaphysics, Morals, Politics, Grammar, Rhetorick, or Logicks).[4.19]

PROMISE AND PERFORMANCE:
EARLY FAILURES AND LATER SUCCESSES

It is interesting to note that both in France and in England the full activity of the societies, as such, was limited to a relatively short period; by 1690 both were in a serious state of decay and their revival in the eighteenth century was practically a new foundation. Their coming into existence and the general support and the interest they aroused in society at large were an indication that science was at that time felt to be exciting, interesting, and might be profitable. It was this last point that was to give rise to serious difficulties. Francis Bacon, like Roger Bacon four centuries earlier, had clearly grasped the idea that the understanding of Nature was the only means of controlling it to the profit of man. But there is a great deal of difference between an idea and an achievement. In fact, it was in one realm alone, though a very important one – that of astronomy and navigation – that the new science, which was practically confined to mathematics and physics, was able to be of real use. Sir Antony Deane did manage in 1666 to find the draught of a ship before it was launched, but this did not notably affect shipbuilding practice. The early Royal Society promised far more than it could perform, and there was some justification, in the short run, for the ridicule with which it was met by the non-scientific intelligentsia, of which the most famous example is the satire by Swift in *Gulliver's Travels*.

In the long run, however, the effect was to be very different. By stimulating the 'naturalist's insight into trades' (p. 461) it was enabled to lay the foundations of that rational evaluation and reconstruction of the traditional arts and manufactures that was to become the Industrial Revolution of the next century. Indeed its work was to lead directly to the central feature of that revolution – the steam-engine, which has every right to be called a *philosophical engine*. It is the fruit not of the work of one or other isolated inventor, but of groups of scientists in the Accademia del Cimento, the Royal Society, and the French Academy (pp. 576 f.).

131. The Royal Society of London for Improving Natural Knowledge was formally founded in 1660, receiving its first Royal Charter two years later. Thomas Sprat (1636–1713), later Bishop of Rochester, wrote the Society's first history in 1667, although it was not so much a history as a defence of the 'new philosophy'. This frontispiece shows Lord Brouncker, the Society's first president (left), Charles II, its patron (centre), and on the right, Francis Bacon. The gallows seen in the background is an irrelevant and irreverent later manuscript addition.

SCIENCE BECOMES AN INSTITUTION

The foundation of the early scientific societies had another and more permanent effect: it made science into an institution, an institution with the insignia, the solemnity, and with, unfortunately, a certain amount of the pomp and pedantry of the older institutions of law and medicine. These societies became in effect a jury for science, a jury sufficiently authoritative to exclude many of the charlatans and madmen whom the general public found it so difficult to distinguish from genuine scientists; but also, unfortunately, able to exclude, for a time at least, many revolutionary ideas from official science itself (p. 587). The range of interest of the associated scientists of the latter seventeenth century, as their *Philosophical Transactions* show, covered almost every aspect of Nature and of practical life from the distances of the stars to the animalcules in pepper water, from the art of dyeing to the bills of mortality.[4.139; 4.140]

The first manifesto of the newly organized science was the *History of the Royal Society* written in 1667, when it was only five years old, by Bishop Sprat. Inevitably it is more than a history, rather a programme and a defence of experimental philosophy. After denouncing varieties of dogmatic philosophers he approves:

The *Third* sort of *new Philosophers*, have been those, who have not onely disagreed from the *Antients*, but have also propos'd to themselves the right course of slow, and sure *Experimenting*: and have prosecuted it as far, as the shortness of their own Lives, or the multiplicity of their other affairs, or the narrowness of their Fortunes, have given them leave.

He defends the inclusion in the Society of men of all ranks and occupations, and from all countries, and then touches on the essential *raison d'être*, which is:

the temper of *the age wherein we live*. For now the Genius of *Experimenting* is so much dispers'd that even in this *Nation*, if there were one, or two more such *Assemblies* settled; there could not be wanting able men enough, to carry them on. All places and corners are now busie, and warm about this Work: and we find many Noble Rarities to be every day given in, not onely by the hands of Learned and profess'd Philosophers; but from the Shops of *Mechanicks*; from the Voyages of *Merchants*; from the Ploughs of *Husbandmen*; from the Sports, the Fishponds, the Parks, the Gardens of *Gentlemen*; the doubt therefore will onely touch *future Ages*. And even for them too, we may securely promise; that they will not, for a long time, be barren of a Race of inquisitive minds, when the way is now so plainly trac'd out before them; when they should have tasted of these first Fruits, and have been excited by this Example.

He concludes his discussion of the experiments and instruments of the Society by commenting on 'the manner of their Discourse' and the need to remove 'the luxury and redundance of speech'. For this reason they rigorously:

> . . . reject all the amplifications, digressions, and swellings of style: to return back to the primitive purity, and shortness, when men deliver'd so many *things*, almost in an equal number of *words*. They have exacted from all their members, a close, naked natural way of speaking; positive expressions, clear senses; a native easiness: bringing all things as near the Mathematical plainness, as they can: and preferring the language of Artizans, Countrymen, and Merchants, before that, of Wits, or Scholars.

The fact remains that the style of the English language was drastically simplified in the later seventeenth century.[4.96; 4.97] It is a curious commentary on this that 100 years later Samuel Johnson wrote of Sprat:

> This is one of the few books which selection of sentiment and elegance of diction have been able to preserve, though written upon a subject flux and transitory. The History of the Royal Society is now read, not with the wish to know what they were then doing, but how their transactions are exhibited by Sprat.[4.94]

CENTRES OF INTEREST IN TECHNIQUE

It appeared at first that anything and everything could be improved by philosophical inquiry. Nevertheless, certain fields of interest drew the special attention of the *virtuosi*. They were those where the themes of the new philosophy met the most clearly felt needs of expanding trade and manufacture. Foremost among these was the refinement of astronomy as an essential need of ocean navigation, particularly in solving the problem of the longitude. This was indissolubly linked with the problem of the true constitution and working of the solar system, by now accepted but not physically explained. Further, it was astronomy that provided the best field for the new *mathematical* explanation of the universe. The solution finally arrived at by Newton was taken to be, and rightly so, the major triumph of the new science.

But this contemporary interest should not be allowed to overshadow other developments which were in the long run to prove at least as important. One of these was *optics* and the *theory of light*, closely linked by the telescope to astronomy, and by the microscope to biology. Another was *pneumatics*, where the techniques developed in connexion with the vacuum were ultimately to have such enormous industrial importance. The question of the *vacuum* was also the centre of a philosophical controversy going back to the Greeks. The new experimental proofs of its

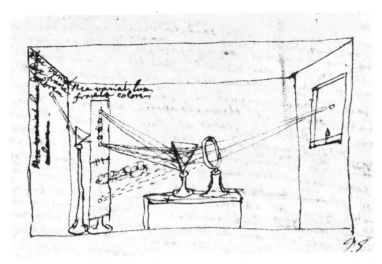

132. Newton's drawing of his prism experiment. Sunlight enters through a hole in the blind on the right and, after the beam passes through a lens, it is dispersed by a prism into a coloured spectrum. A second prism (left) fails to cause further dispersion, thus demonstrating that the colours of the spectrum are the primary components of white light (sunlight). From the New College, Oxford, Collection at the Bodleian Library.

existence helped to revive the *atomic* hypothesis of Democritus. The revived atomic or corpuscular theory provided a first clue to rational and quantitative explanations in the field of chemistry, which had hitherto been one of technical recipes and mythical explanations. Chemistry, in turn, was linked with the beginnings of *physiology*. Questions on the nature of the blood, the function of the lungs, the action of nerves and muscles, and the process of digestion, were all discussed and experimented on in the spirit of the new materialist philosophy. This range of subjects was not beyond the reach of individual men of the time, and indeed it is best illustrated in their lives and works. Outstanding among them were Robert Boyle and his one-time assistant Robert Hooke.

ROBERT BOYLE

The Hon. Robert Boyle was born at Lismore in 1627, the seventh son and thirteenth child of Richard Boyle, first Earl of Cork, a ferocious and successful land-grabber of Elizabethan times.[4.36; 4.104; 4.122] He spent his

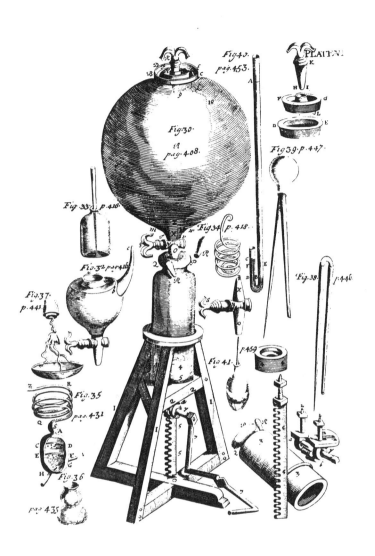

133. Vacuum pump constructed for Robert Boyle by Robert Hooke. First published in Robert Boyle, *New Experiments Physico–Mechanicall, Touching the Spring of the Air*, London, 1660. From Peter Shaw, *The Philosophical Works of the Hon. Robert Boyle, Esq.*, London, 1725.

most impressionable years in the Puritan atmosphere of Geneva, where he underwent a religious conversion, like his contemporaries Pascal* and Steno. Unlike Pascal, however, this did not turn him against science but made him strive to use it in the support of revelation. Partly for this reason and partly because he was a lifelong invalid, he led an ascetic life, took no sides in the Civil War, and devoted himself and his considerable wealth to the pursuit of the new experimental philosophy. He worked with the 'Invisible College' in Oxford and was one of the first promoters of the Royal Society, of which he was offered the presidency in 1680 but declined on account of a scruple about the oath. Boyle was, indeed, the central figure of the early days of the Royal Society, as Newton was of its prime. He wrote profusely on religious and scientific subjects. His most famous works, apart from that on the *Spring of Air*, were the *Seraphick Lover*, *The Skeptical Chymist*, and the *Unsuccessfulness of Experiments*. His early interest in the atomic theory led him to his epoch-making work on the vacuum and gas laws. After that he was not so successful, partly because he lacked adequate mathematical and experimental skill, but mostly because he attempted to explain problems in chemistry by mechanical theories which could not be applied to them and before enough facts had been accumulated to solve them by any other means. His interests ranged further still into physiology and medicine where there was still less hope of solid achievement. Nevertheless he infected others with his interests and enthusiasms and much of the success of science in the next century was due to his inspiration. In Boyle we can see the pietistic and philanthropic aspects of the new science. He combined the desire to show the glory of God revealed in His works with that of helping his fellow men, and he actually joined the boards of the Bermuda and East India Companies to further his schemes for converting the heathen. Nevertheless in the achievements of these ends he was, unlike the medieval pietists, intensely practical. In his pamphlet *That the Goods of Mankind May be Increased by the Naturalist's Insight into Trades* [4,40] he wrote:

I shall conclude this, by observing to you, that as you are, I hope, satisfied, that experimental philosophy may not only itself be advanced by an inspection into trades, but may advance them too; so the happy influence it may have on them is none of the least ways, by which the naturalist may make it useful to promote the empire of man. For that the due management of divers trades is manifestly of concern to the publick, may appear by those many of our English statute laws yet in force, for the regulating of the trades of tanners, brick burners, and divers other mechanical professions, in which the lawgivers have not scorned to descend to set down very particular rules and instructions.

ROBERT HOOKE

In many ways Boyle contrasts with his first assistant and lifelong friend, Robert Hooke. If one was a nobleman condescending to science, the other was a poor man who had to make his living out of science while he pursued it. The son of a clergyman in the Isle of Wight, Hooke managed to secure a servitorship at Oriel College at the time when Boyle had come to Oxford. He early attached himself to him and, in fact, probably made all his apparatus and carried out most of his experiments on the vacuum and gases. Boyle certainly did not shine as an experimenter after Hooke left him. Hooke was made curator of experiments of the Royal Society when it was founded, and as well as carrying out his heavy duties managed to supplement his meagre and irregular salary by being largely responsible for the plans of the new City of London after the Great Fire of 1666.

If he had been in a more secure social position and had not suffered from his ugliness and chronic ill health, he would not have been the difficult, suspicious, and cantankerous person he was, and his quite decisive role in the history of science would have been fully recognized. If Boyle was the spirit behind the Royal Society, Hooke provided it with eyes and hands. He was the greatest experimental physicist before Faraday, and, like him, lacked the mathematical ability of Newton and Maxwell. His interests ranged over the whole of mechanics, physics, chemistry, and biology. He studied elasticity and discovered what is known as Hooke's law, the shortest in physics: *ut tensio sic vis* (extension is proportional to force); he invented the balance wheel, the use of which made possible accurate watches and chronometers; he wrote *Micrographia*, the first systematic account of the microscopic world, including the discovery of cells; he introduced the telescope into astronomic measurement and invented the micrometer; and he shares with Papin the credit of preparing the way to the steam-engine.

Probably his greatest contribution to science is only now beginning to be recognized: his claim to have originated the idea of the inverse square law and universal gravity. Here, as we shall see, he was outclassed by the superb mathematical achievement of Newton, but it now seems that the basic physical ideas were Hooke's and that he was quite unjustly robbed of the credit for them (pp. 480 f.). Hooke's life illustrates both the opportunities and the difficulties that the gifted experimenter could find in the seventeenth century. It also brings out the enormous store of inventiveness and scientific insight that had lain concealed for thousands of years in the brains and hands of natural craftsmen.[4.63]

7.8 Making the New World-Picture

The accent of the period was one of *extensive* inquiry covering the
whole field of Nature and the arts, and *constructive* theory in those parts
where mathematical methods could be applied. It was no longer necessary, as it had been in the previous period, to concentrate on upsetting the
physics of Aristotle or the physiology of Galen. The theories of Copernicus, Galileo, and Harvey were almost unanimously accepted by the new
virtuosi. Where they differed from their predecessors was in attempting
to give them a deeper physical and philosophic meaning. First in the
field was the system of Descartes with its emphasis on mere extension
and the complete and continuous filling of the universe with subtle
matter acting by impulsion from one part to another. It was the doctrine
of the *plenum*.

THE CORPUSCULAR PHILOSOPHY : GASSENDI

But there was another view, and a far older one, beginning to make itself
felt. The attack on Aristotle left the way clear for Democritus and his
atomic theory (pp. 183 f.). This was brought to the notice of the scientific
world by a learned and penetrating mathematician and philosopher,
Gassendi (1592–1655), a Provençal priest. If he had not been of a modest
and retiring nature he would not have been so easily overshadowed by
his contemporary, Descartes; for his influence on science was great. He
was an astronomer of note – he was the first to observe the transit of
Mercury – and one of the founders of meteorology, being the first to
study parhelia (mock suns) and the aurora borealis. He did far more than
resurrect the old atomic theories as set out by Epicurus and Lucretius;
he turned them into a doctrine which included the Renaissance advance
in physics. Gassendi's *atoms* were massive particles with inertia and they
moved in the *vacuum* which Galileo's successors had proved to exist.
His definition of atoms is that given, almost word for word, by Newton
in his *Opticks* fifty years later. He put this view forward so persuasively
that it was accepted, almost without their noticing it, by all those natural
philosophers who had not sworn adherence to Descartes' plenum, with
its vortices.

The *corpuscular hypothesis* was obviously well suited to the mathematical–mechanical bent of the time. Following the dynamics of Galileo
and Descartes it was far easier to work out the motions of such small
point-like particles than of a piece of homogeneous space. Thanks to

Gassendi's piety the atoms were also purged of their atheistic and sub-versive associations (p. 184). He made explicit the implications of the new mechanics by demanding of God not the continuous operation of the material world, but only an impulse given to all the atoms at the beginning of time which should determine by divine providence all their future movements and combinations.

PHILOSOPHICAL INSTRUMENTS: OPTIC GLASSES

The emphasis on experiment in the new science implies the use of appara-tus and particularly of instruments made especially for the purpose. Nevertheless the material equipment of the new scientists was still of the simplest. Only telescopes had to be large and expensive. Almost any house could be filled with an *elaboratory* (or glorified workroom) which might hold a furnace with a few retorts and still heads, a balance, a micro-scope, and some dissecting instruments, one of the new air-pumps, a thermometer, and a barometer. Anything else was improvised. With such equipment the greatest discoveries in all branches of science could be made. It will be convenient to treat those in optics, pneumatics, chemistry, and physiology before passing on to the central theme of the mechanics of the heavens.

It was the practical and accidental discovery of the telescope at the beginning of the century that gave rise to a new interest in optics; for once an instrument exists the need to improve it leads to searching for explana-tions of how it works. In attempting to do so scientific principles leading to other instruments are discovered. Seventeenth-century optics grew largely out of the attempt to understand the nature of refraction, on which the telescope is based, and to remove the defects which it was soon observed to have.

On the first problem of the nature of refraction they had to start where Alhazen (p. 278) and his medieval followers, Dietrich of Freiburg and Witelo (p. 302), had left off 400 years before. They had established that rays were bent or broken – refracted – on meeting a denser medium. But they could not find the law of refraction and therefore could not calculate the action of a lens. The Dutchman, Snell (1591–1626), found the correct law, which Descartes appropriated and explained in terms of moving particles of light which needs must travel faster in the refracting body than the air, an unlikely conclusion which was to lead to much confusion later. With Snell's law, optics seemed to become part of geometry and it should have been possible to construct perfect telescopes. Actual tele-scopes, however, remained irritatingly imperfect. In particular, images of stars were seen surrounded by coloured haloes. That light passing

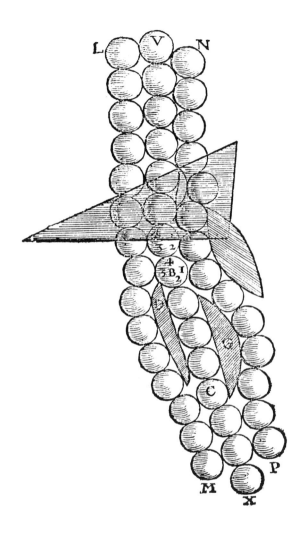

134. It was Descartes who first published the law of refraction of light, discovered by Willebrord Snell (1591–1626) in his *La Dioptrique*, 1637. Descartes' explanation of the nature of light and of the mechanism of refraction was in error. From the first Latin edition of Descartes' correspondence, 1668 (see plate 129). The picture shows light particles impinging on a prism and being little disturbed by lenses lying in their subsequent path.

through transparent bodies emerged with the colours of the rainbow had long been known. With the object of elucidating the rainbow the medieval scientists even carried out extensive experiments on prisms, but had got no further than noting the fact that red light was least and blue was most refracted.[4.52] Descartes in his study of the rainbow was not able to improve on them. The solution of the problem of colour was only to be found by Newton, and was his first recognized achievement in physics. (His career will be treated later, pp. 481 f., in conjunction with his work on gravitation.)

NEWTON'S 'OPTICKS': THE DOCTRINE OF COLOURS

Newton had first tried to avoid the difficulty of coloured images by doing without the refraction which caused it. He built the first reflecting telescope, the prototype of the giants of today and also of the more recent device, the reflecting microscope. Not satisfied with this he attacked the problem of colours directly, taking up the experiments of Descartes on the prism where he had left them. By a most brilliant combination of experimental technique and logic he was able to show that the colours of the prism, or of the rainbow, are not created by it but are the intrinsic components of ordinary white light. His researches, however, did not help him to solve his original problem; indeed he was able to show to his own dissatisfaction that it was impossible to correct the dispersive or colour-making properties of lenses. In this he was wrong, and his authority held up the practical development of telescopes for about eighty years. A Swedish mathematician, Klingenstjerna (1698–1765), seems to have been the first to repeat Newton's experiments with sufficient care to show his error. It was not until 1758 that Dollond, the instrument maker, hearing of Klingenstjerna's work, was able to use the idea of balancing two kinds of glass of different refractivities and dispersion against each other, so producing the achromatic lens which is the basis of all modern optical instruments.

LIGHT AS PARTICLES OR WAVES: HUYGENS

Newton, in his optical studies, considered kinds of colour other than those of the rainbow, notably those produced by reflection from thin layers, such as oil on water. It was there he found the first hint of the discontinuity or 'graininess' of both matter and light. This strengthened his conviction, already gained by philosophical inclination and mathematical convenience, that matter is atomic. Unfortunately the same conviction caused him to follow Descartes and treat light as atomic, its rays being the trajectories of particles reflected just as a ball bounces off a wall. Other phenomena producing colours pointed to a different conclusion. Grimaldi (1618–63) had long before Newton studied the colours found at the edges of shadows, particularly those of fine slits or hairs. He also found that the rays of light were not quite straight but slightly bent – diffracted – on passing near an object. He put both phenomena down to waves, like the familiar ripples in water, or pulses of sound, with the different colours having different wave-lengths like the notes of music.

Huygens developed this idea mathematically and showed how the *wave theory of light* would account both for diffraction and the colours of thin plates. Moreover he explained, far better than Newton had, the curious property of Iceland spar (calcite) of showing objects seen through it as double. Here again, however, Newton's superior authority carried the day and the wave theory of light had to wait more than a century before it was rehabilitated (pp. 575, 610 f.).

THE MICROSCOPE: THE NEW WORLD OF SMALL THINGS

Just as the telescope in the hand of Galileo had found the secret of the stars, that other optic glass, the microscope, in the hands of a number of seventeenth-century observers such as Malpighi, Hooke, Swammerdam (1637–80), and the incomparable Dutch clothier, Leeuwenhoek (1632–1723), opened up a new world of the very small.[4.67] Insects, the parts of

135. A reproduction of a Leeuwenhoek microscope. The lens is the round disk on the flat metal plate, and in front of it is a pin on a screwed rod: the object under investigation was placed on this and the eye was held close to the lens on the (here) left-hand side of the flat plate.

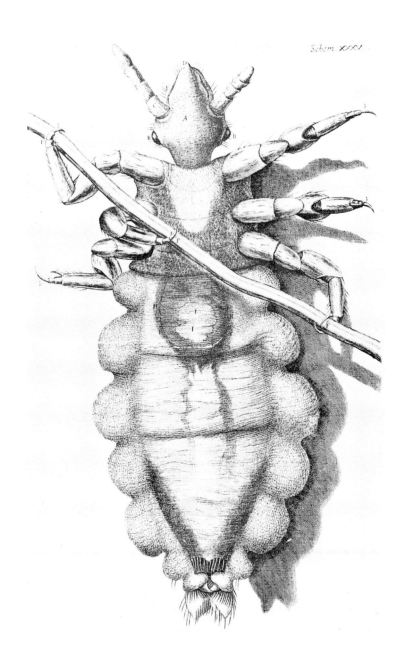

136. Robert Hooke's drawing of a louse, '. . . a Creature so officious that 'twill be known to every one at one time or other. . . '. From his *Micrographia; or some Physiological Descriptions of Minute Bodies made by Magnifying Glasses with Observations and Inquiries thereupon*, London, 1665. Christopher Wren helped Hooke to prepare the plates for this book.

plants, the small creatures that live in water, even the minute bacteria and the spermatozoa that carry the principle of generation, were all observed and became the objects of wonder, speculation, and argument. The anatomy of larger animals was also refined and Harvey's theory of the circulation of the blood fully confirmed. But whereas the telescope, whether nautical or astronomical, had from the very start a real, practical use, the microscope did not prove its value until 200 years later, in the hands of Koch and Pasteur, for combating bacterial disease. Largely for this reason these early microscopical studies did not immediately lead to any great development either of microscopy or biology; what was seen remained more amusing and instructive – in the philosophical sense – than of scientific or practical value.

THE VACUUM AND THE BAROMETER

The development of *pneumatics* far beyond the bounds reached by the Greeks (pp. 220 f.) was the first great step forward in physics that was to lead rather to industrial than to astronomical and navigational consequences. The decisive discovery that brought it about, the actual production of the *vacuum*, was itself derived directly from practical hydraulics. Hitherto the existence of the vacuum had been a philosophic question to be settled by argument (pp. 183 f.); from 1643 it was to become a matter of practical demonstration. Galileo was, in his latter days, concerned with the reason why it was impossible to raise water by ordinary suction pumps more than about thirty-two feet. This fact, long known to miners and well-sinkers, had not hitherto attracted the attention of the learned. Galileo attributed it to the inability of a water column to bear its own weight, though he could find no satisfactory explanation why once it had broken it did not fall right down, and attributed this to a limited *horror vacui*.

It was not until the year after Galileo's death that his pupil Torricelli (1608–47) had the ingenious idea of using mercury instead of water, and was thus able to work with a column of manageable height, for in the inverted tube the mercury would not rise above thirty inches, thus giving the same pressure as the water column of fifteen pounds per square inch. He had the intellectual courage to see that the real explanation was that the pressure of the air held up the column of mercury, so that the instrument was a *barometer*, a means of measuring the weight of the atmosphere. The space at the top of the column was the real *vacuum* which Nature was supposed to abhor. Indeed, as we have seen (pp. 202 f.), Aristotle had already proved a vacuum to be impossible because air, opening in front and closing in behind, was needed for violent motion. Its discovery was a last and fatal blow to Aristotle's mechanics, though every effort was made to deny it or explain it away. Torricelli's explanation was, however, soon confirmed by Pascal's (1623–62) experiment of taking a barometer up a mountain and noticing the fall in pressure.

VON GUERICKE'S AIR-PUMP

The story was then carried forward by a remarkable character, the prototype of the heavily endowed scientists of today, Otto von Guericke (1602–86), the mayor of Magdeburg and ex-Quartermaster of Gustavus Adolphus, and a man of considerable means and great enterprise. Von Guericke did things in a big way; he spent £4,000, a stupendous sum in those days, on his experiments. He first of all tried to produce a vacuum by the straightforward method of pumping the water out of a closed barrel. The barrel burst, so he made a stronger vessel of brass. Afterwards he devised an air-pump and succeeded in producing vacua in large vessels. One of these he used for his celebrated experiment when sixteen horses a side were needed to pull apart two hemispherical vessels in the presence of the Emperor and his court. The Magdeburg hemispheres furnished a most impressive demonstration of the material truth of the new science. But the experiment did more than that; it showed people that the vacuum of the pressure of the air was a most powerful force that only needed ingenuity to harness it for useful purposes. Von Guericke himself thought of transferring power through evacuated tubes, an idea that was afterwards developed in the vacuum brake for the railways.

Von Guericke's pumps were much improved by Boyle, or more probably by Hooke, who was then in his pay. With this pump Boyle demonstrated many new and strange effects. He showed, for instance, that without air, sound could not travel, but light and magnetism were not affected. He also found what might have been expected, but was nevertheless a striking demonstration, that both life and combustion were impossible in a vacuum, and thus provided the first clues to the great chemical and physiological revolution of the next century.

The use of the air-pump, particularly the effort involved in pumping, led Boyle to a study of the behaviour of air, both compressed and ex-

137. The chemical laboratory of Ambrose Godfrey, often known as 'Mr Godfrey'. Godfrey became Boyle's operator in about 1683 and built his own laboratory in about 1706. The equipment was so outstanding that any visitor to London interested in science visited Southampton Street where Godfrey had set up as an apothecary. From *The Complete Dictionary of Arts and Sciences*, edited by Temple Henry Croker, Thomas Williams, and Samuel Clark, London, 1764.

panded. Thus he discovered the first scientific law outside that of simple mechanics, what he called the 'Spring of Air', the law that we now know as Boyle's Law: that the pressure multiplied by the volume of a certain amount of air is constant – or rather, as was found later, is directly proportional to the temperature.

The idea of using new natural forces to satisfy human needs had never entirely died out, and was bound to come up in an era of scientific enterprise such as the seventeenth century, when there was a mounting need of brute force to pump the mines and set going the wheels of flourishing industry. One obvious force to use was that of fire, especially ever since the power of fire was made manifest in the cannon. One of the first crude ideas was an internal-combustion engine using gunpowder where we use petrol. After that inventors turned to the expansive power of steam. These direct methods were bound to fail, not because they were intrinsically wrong, but because the technique of the time did not provide vessels strong enough to deal with pressures of this magnitude. Denis Papin (1647–1712), Huygens' assistant, who afterwards worked for a while with Boyle, did manage to make a *digester* in which he reduced bones to soup, but his *pressure cooker* has only come into use in our own day. He also took the first steps to a practical steam-engine. The

138. The famous demonstration of the power of a vacuum using two teams of horses, each harnessed to one half of a 'Magdeburg' sphere. From *Experimenta Nova* by Otto von Guericke (1602–86), of Magdeburg, published at Amsterdam in 1672.

way to the use of steam power was to lead through the vacuum, as will be shown in the next chapter.

THE FALSE DAWN OF RATIONAL CHEMISTRY

The discovery of the vacuum furnished the first clue that might have led to the development of rational chemistry in the seventeenth century instead of 100 years later. The vacuum pump showed how air was necessary both to combustion and breathing, and centred interest on the twin problems of flame and life. Boyle, Hooke, and Mayow, following a clue dropped by Paracelsus, almost succeeded in proving that air contained something that was essential to burning and that turned arterial blood red. Boyle referred to it as 'a little vital quintessence (if I may so call it) that serves to the refreshment and restauration of our vital spirits'. Mayow called it 'Nitro aerial spirit' – thus linking it with gunpowder – which was to become Lavoisier's *oxygen*. But they got no further for two fundamental reasons: lack of suitable scientific theory and inadequate techniques and materials.

Chemistry had never been part of the classical canon, and the Aristotelian elements, earth, water, air, and fire, had always had a meteorological and physical aspect, rather than a chemical one (p. 176). Arab and medieval chemistry, or rather alchemy, was, however, thoroughly mixed with an astrology that linked the metals with the planets. The collapse of the Aristotelian–Platonic world-picture meant that chemistry without its airs and planetary influences had no intellectual basis left, as Boyle pointed out in *The Skeptical Chymist*. Nor did the Arabic–Paracelsan 'spagyric' chemistry of the three principles – mercury, sulphur, and salt – fare any better (p. 398). The principles were far too vague and changeable to fit into a corpuscular philosophy which was specifically designed to exclude *occult qualities*. Boyle himself managed to give a precise definition of an element, though a negative one:

No body is a true principle or element . . . which is not perfectly homogeneous but is further resolvable into any number of distinct substances how small soever.

Unfortunately, the technique of chemistry could give no guarantee, apart from a few metals, as to what was an element; and Boyle's criterion was inapplicable for another hundred years. He recognized this himself in his essay *On the Unsuccessfulness of Experiments*.

Newton, who worked at chemistry for much longer than at physics, got no further in practice.[4.129] In theory he had evolved, as Vavilov[4.159] has shown, a picture of the atom composed of shell within shell of parts held together successively more firmly. This was a striking and quite logical anticipation of the modern atom with its electrons and nuclei, but it lay forgotten for nearly 300 years. In the seventeenth century chemistry was not yet in a state in which the corpuscular analysis could be applied. For that it needed the steady accumulation of new experimental facts that was to come in the next century. Chemistry, unlike physics, demands a multiplicity of experience and does not contain self-evident principles. Without principles it must remain an 'occult' science depending on real but inexplicable mysteries.

As long as chemistry revolved around the same materials as were known to the Ancients, it tended to become stereotyped. But after the fifteenth century the chemical world expanded rapidly. New substances with remarkable properties, such as phosphorus, were accidentally produced, and new metals such as bismuth and platinum were discovered in the Old and New Worlds. To explain their properties new theories, continuously being checked by new practice, were needed. They were necessarily, at first, qualitative and obscure, but they formed an essential foundation to more precise theories. All the time, in response to the demands of an ever more specialized trade and industry, there came the need for particular chemicals – saltpetre, alum, copperas (iron sulphate), oil of vitriol (sulphuric acid), soda – which gave birth to a chemical industry from the experience and the problems of which was to come the rational chemistry of a later age.

SEVENTEENTH-CENTURY BIOLOGY

The world of living things, with its enormously greater complication, was bound to be far more difficult to explain than that of chemical transformation. It was therefore not surprising that the new mechanical, corpuscular philosophy, in spite of its pretensions, was of little real service. Sanctorius (1561–1636) weighed himself in a balance while eating and sleeping but could not explain the changes he observed. Descartes' idea of the animal-machine and the man-machine which differed only by the attachment of a rational soul steering it through the pineal gland did little to advance physiology. Borelli (1608–78) pushed the analogy further and accounted, on mechanical principles, for the limb movements of men and animals. Hydraulics had worked well for the heart and blood but were of little use for the brain and the nervous fluid.

Where the seventeenth century did make a critical advance was in observation, particularly in using the microscope (pp. 467 f.), which revealed for the first time the spermatozoa responsible for generation. More immediately important was the work of Nehemiah Grew (1641–1712), who laid the foundations of plant physiology, and of John Ray (1627–1705), a blacksmith's son, who took the first steps towards a scientific classification of plants and, less successfully, of animals.

The biological investigations of the late seventeenth century were, in practice, of little immediate use to agriculture. The changes that were made, and they were great, particularly in horticulture, were due rather to the careful and slow improvements of traditional practice under exceptionally favourable economic conditions. It was in Flanders and Holland that it was possible to find men of substance able and willing to put *capital* in the form of implements and manure into their farms and at the same time to be assured of an ample and growing market for the improved produce. Holland was the nursery from which the new methods, thanks to the work of enthusiastic amateurs like John Evelyn (1620–1706), were to pass to England (p. 412, *n*. p. 380).

The direct method of observation and experiment was to be more immediately fruitful in medicine, though progress was disappointingly slow. The idea that medicine was a science to be discovered from the study of patients rather than a doctrine to be practised on them, though as old as Hippocrates, had been largely forgotten. It was renewed in this time by doctors like Sydenham (1624–89), who, besides being a great clinician, was in touch with all the science of his time.

7.9 Celestial Mechanics: The Newtonian Synthesis

While all these achievements bear witness to the great flowering of scientific activity in many fields, the central interest and the greatest scientific triumph of the seventeenth century was undoubtedly the completion of a general system of *mechanics* capable of accounting for the motion of the stars in terms of the observable behaviour of matter on earth. Here the moderns were in effect settling their accounts once and for all with the ancient Greeks. Ancients and moderns were both agreed on the importance of the study of the heavens. But because the interests of the latter were now more practical than philosophic, they required a very different kind of answer. Finding that answer in a complete and satisfying form was the work of a sequence of mathematicians and astronomers, including almost all the great names of science of the period – Galileo, Kepler, Descartes, Borelli, Hooke, Huygens, Halley, Wren – but all was to lead up to the clear unification of mechanics in Newton's *De Philosophiae Naturalis Principia Mathematica*, where he set out and proved his theory of universal gravitation.

The intrinsic interest of the problem of the movements of the solar system was still very great, though, in fact, its philosophical and theological significance had already vanished with the destruction of the cosmology of the Ancients. The trial of Galileo was indeed in the nature of a futile parting shot by clerical Aristotelianism. But the new edifice that was to take its place would not be complete unless an acceptable physical explanation of the system of Copernicus and Kepler could be found. That was one reason why almost every natural philosopher speculated, experimented, and calculated with the aim of finding this explanation. Some got very close to it, particularly Hooke,[4.63] until Newton's success ended the chase.

FINDING THE LONGITUDE

The astronomers had another and even more compelling reason for discovering the laws of motion of the solar system. This was the need for

139. The title page of the first edition of Newton's *De Philosophiae Naturalis Principia Mathematica*, London, 1687, bearing Samuel Pepys' imprimatur as President of the Royal Society. This book has exerted more influence on thinking in natural science than any other work.

PHILOSOPHIÆ

NATURALIS

PRINCIPIA

MATHEMATICA.

Autore *J S. NEWTON*, *Trin. Coll. Cantab. Soc.* Mathefeos Profeffore *Lucafiano*, & Societatis Regalis Sodali.

IMPRIMATUR·

S. P E P Y S, *Reg. Soc.* P R Æ S E S.

Julii 5. 1686.

L O N D I N I,

Juffu *Societatis Regiæ* ac Typis *Jofephi Streater*. Proftat apud plures Bibliopolas. *Anno* MDCLXXXVII.

140. Giovanni Domenico (or Jean Dominique) Cassini (1625–1712), the first of three generations of Cassinis to direct the Paris Observatory which was founded in 1672. The observatory building is in the background, and in front of it is a long focus refracting telescope. From an early eighteenth-century copperplate engraving.

astronomical tables far more accurate than had sufficed in the days when astronomy was required mainly for astrological prediction. The needs of navigation were far more stringent. The determination of a ship's position at sea, and particularly the more difficult part of the position, the longitude, was a recurring problem. It became more and more urgent as a larger and larger share of the economic and military effort of countries was spent in overseas ventures, especially of those countries that were themselves the centres of scientific advance: England, France, and Holland. The finding of the longitude was a question that was to occupy both the learned astronomers and the practical sailors for many decades, even centuries. It was for the purpose of assisting in the solution of this practical problem that the first nationally financed scientific institutions were set up – the Observatoire Royal at Paris in 1672 and the Royal Observatory at Greenwich in 1675.

The question of the determination of longitude is essentially one of determining absolute time – or, as we now would call it, Greenwich time – at any place. This, compared with the local time, gives the time interval which is directly convertible to longitude. At any place there are, or were before the invention of radio, only two methods of determining the Greenwich time: one by observing the movements of the moon among the stars – a clock already fixed in the sky; and the other by carrying around an accurate clock originally set at that time. The first required extremely accurate tables for the prediction of the place of heavenly bodies, the second absolutely reliable clock mechanisms. All through the seventeenth and a large part of the eighteenth centuries both lines of attack were pursued without definite advantage falling to either. There was an immediate stimulus to thought, observation, and experiment in both directions, a stimulus in part simply mercenary but also one of national and individual prestige.

THE CHRONOMETER

The two methods were at first sight quite different: one was concerned with a movement of some material-controlling mechanisms, the other with that of spheres in empty space; but as they were studied both were found to have a common basis in *dynamics*. It was Galileo himself who had discovered that the ideal controller, that beat constant time, was the *pendulum*. Hooke made the essential practical contribution of substituting the spring-controlled balance wheel, which was not upset by the motion of the ship, for the pendulum. In either case accurate time-keeping depended on knowing the laws of motion of bodies in oscillation, and it was here that Huygens solved the problem and laid the basis

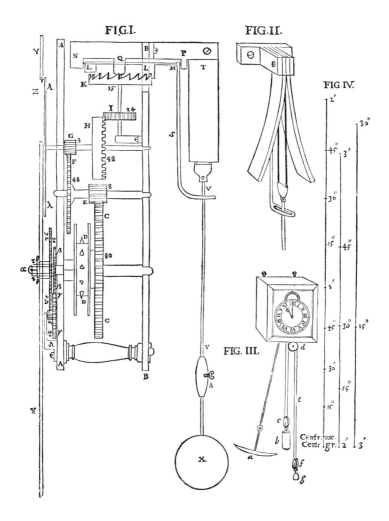

FIG.I.   FIG.II.

FIG.IV.

FIG.III.

141. A page from Christian Huygens' *Horologium Oscillatorium*, Paris, 1673, which gave details and discussed the principles of pendulum clocks and made accurate timekeeping a practical possibility, especially for scientific work. Figure I is a side view of Huygens' clock with a horizontal escapement wheel connected by pallets LL to the pendulum; Figure II his design of cycloidal jaws to provide a wide swing with improved isochronomism; Figure III is the complete clock; Figure IV an analysis of the centre of oscillation and gravity of his cycloidal pendulum.

of the first chronometer, as set out in his book *Horologium Oscillatorium* (1673). But a long time had to pass before these principles could be turned into effective practice through improved workmanship, and Harrison's chronometer could in 1765 finally win the prize offered by the Admiralty for achieving the longitude.*

PLANETARY MOTIONS : THE DOCTRINE OF ATTRACTION

It was, however, the purely astronomical approach that, though it failed to provide the practical solution, was to prove far the more valuable to the science of the future. This was because of the stimulus it gave to the finding of a mathematical and dynamical solution to the problem of planetary motion. Many people had speculated as to why the planets should move round the sun in the orbits which Kepler had first shown were elliptical; they had even guessed that they might be held there by some force of attraction. In fact the idea of attraction had been a common one ever since Gilbert's study of the magnet (pp. 434 ff.), and even before. The magnet showed that attraction was possible at a distance and Gilbert himself had suggested that what held the planets in their position and indeed drove them round their orbits might just be magnetism.

Borelli in 1666 introduced the important idea that the movements of planets implied the existence of the need to balance the centrifugal force, such as that exerted by a stone in a sling, by some other force which he characterized as the force of gravity extending beyond the immediate neighbourhood of the earth to the moon and from the sun to the planets. To account for an elliptical orbit, with the planet moving faster as it nears the sun, the force of gravity must increase to balance the increased centrifugal force. The force of gravity is therefore some function of the power of the distance. The question now became: What function? Hooke, who had already suspected that gravity diminished with the distance, tried to confirm it by looking, though in vain, for the variation of weight in a body on the ground, in a mine-shaft, and at the top of a steeple.

The prevailing theory of gravity remained that of Descartes: namely, that heavy bodies were sucked down to their centres of attraction by 'some secret principle of unsociableness of the ethers of their vortices', to quote Newton, who adhered to this theory as late as 1679.[4.120]

Things could go no further until these general ideas could be reduced to a mathematical form and checked with observations. The first step to this was taken by Huygens in 1673, when in connexion with his work on pendulum clocks he enunciated the law of centrifugal force, showing that it varied as the radius and inversely as the square of the period. Now the square of the period, according to Kepler's third law, was proportional to the cube of the radius, and it follows therefore that the gravitational pull or centripetal force to balance the centrifugal force must depend on the radius divided by its cube, that is on the inverse square of the radius. Hooke, Halley, and Wren had made this deduction by 1679. Two problems remained: that of the explanation of elliptic orbits; and the mode of action of large attractive bodies. Hooke wrote to Newton, putting these problems, but received no reply, and in 1684 Halley offered a prize for their solution. It was clear that the answer was very near, but, though many men had led up to it, only one had the mathematical ability to find it and to draw the revolutionary conclusions that followed from it.

ISAAC NEWTON

That man was Isaac Newton, one of the younger generation of Fellows – born in 1642, the year Galileo died – but already well known for his mathematical and optical researches. Newton came from the new rural middle class that had already produced Cromwell and the parliamentary officers. He was the posthumous son of a small Lincolnshire farmer with connexions good enough to send him to Cambridge, where he studied with no particular distinction. In 1663 he came into contact with the learned and travelled Isaac Barrow (1630–77), the new Lucasian professor of mathematics, who appreciated his abilities and got Newton appointed to his chair in 1669 at the age of twenty-six, though he had published nothing and attracted little notice. He remained at Cambridge until at the height of his fame he was, in 1696, appointed Warden, and later Master of the Mint, at £400 a year, a job that he was considered very lucky to get and the duties of which he carried out conscientiously.[4.113]

At Cambridge Newton worked on optics, many other branches of physics, chemistry, biblical chronology, and theology of a heretical, Arian kind. He seems to have had but little influence on the University and never founded a school. It was there that he came under the influence of a deeply religious group of Platonists led by Henry More, and through them Platonic elements entered into his philosophy and consequently into that of modern science.[4.44] In general he conformed to the views of his class, represented Cambridge University in Parliament, and supported the Whig compromise in politics. This helped to make his ideas, which were only later to show their revolutionary potentialities, appear respectable at the start. Newton was personally an extremely odd character, very reserved and retiring, even secretive. He never married and would not accept ordination because of his doubts about the Trinity. He knew

142. Isaac Newton (1642–1727), from a mezzotint executed in 1740 by James Macardel (1710–65), one of the finest British workers in mezzotint, from a portrait by Enoch Seeman.

enough to make him very self-critical; but this made him even more resentful of the criticisms of other people.

Newton's public entry into the discussions of gravitation came late. We know now, from recently discovered manuscripts, that already in 1665, as he claimed in his old age, he had discovered the essential principle of the inverse square law of gravitation, which he had derived from Galileo. Incidentally he had discovered the law of centrifugal force, some ten years before Huygens, to whom it had always been attributed. Why he did not publish his theory of gravitation for nearly twenty years still remains a mystery. Newton was always a perfectionist and it may well be true that discrepancies based on the bad measurements of others deterred him. We have no evidence as to when he was able to calculate the elliptical character of a planet's orbit. This may have been the real sticking point.

When he did publish, drawn into the controversy with Hooke by his friend Halley, Newton's contribution was decisive. It lay in finding the mathematical method for converting physical principles into quantitatively calculable results confirmable by observation, and conversely arriving at the physical principles from such observations. In his own words from the preface of *Principia*:

> I offer this work as the mathematical principles of philosophy, for the whole burden of philosophy seems to consist in this – from the phenomena of motions to investigate the forces of Nature, and then from these forces to demonstrate the other phenomena; ... I wish we could derive the rest of the phenomena of Nature by the same kind of reasoning from mechanical principles, for I am induced by many reasons to suspect that they may all depend upon certain forces by which the particles of bodies, by some causes hitherto unknown, are either mutually impelled towards one another, and cohere in regular figures, or are repelled and recede from one another. These forces being unknown, philosophers have hitherto attempted the search of Nature in vain; but I hope the principles here laid down will afford some light either to this or some truer method of philosophy.

THE INFINITESIMAL CALCULUS

The instrument by which he did this was the infinitesimal calculus or, as he called it, the method of fluxions (the even flowing of a continuous function). This marked the culmination of the work of many generations of mathematicians, from Babylonian predecessors through Eudoxus and Archimedes (p. 185). In the seventeenth century it was rapidly developed through the work of Fermat and Descartes. It was put in the form we know it by Leibniz (1646–1716) (p. 518). Whether Newton or Leibniz deserves the greater credit for it – a subject of bitter controversy at the time – is not, from the point of view of the progress of science, of any great moment. What is important is that Newton used his calculus to solve vital questions in physics and taught others to do the same.

By its use it is possible to find the position of a body at any time by a knowledge of the relations between that position and its velocity or rate of change of velocity at any other time. In other words, once the law of force is known, the path can be calculated. Applied inversely, Newton's law of gravitational force follows directly from Kepler's law of motion. Mathematically they are two different ways of saying the same thing;

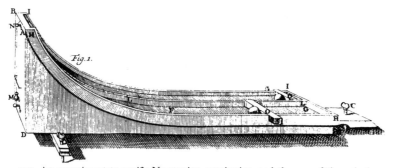

143. An experiment to verify Newtonian gravitation and the use of the calculus. Newton used the calculus in working out his famous theory of universal gravitation and applying it to falling bodies on Earth and the curved orbits of planets under the Sun's gravitational influence, even though, when he wrote the *Principia*, he used the geometrical methods more familiar to his readers. Newton's friend, the Dutchman Willem 'sGravesande, published his *Physices Elementa Mathematica* in two volumes in 1720, in which he gave experiments to demonstrate the truth of Newton's hypothesis. This engraving, from the second edition of 1747, shows a demonstration of the relationship between the fall of a terrestrial body and the 'fall' of the Moon towards the Earth in its orbit which gives its orbit the required curvature. The curved wooden surface represents the orbit: two balls are released from different heights so that they end up at different positions (o, o) at the same instant. The demonstration allows the relationship to be established, since the curved sides of the slope are graduated and the distances travelled in a curved path may be compared with the vertical fall and the amount of the acceleration. The calculus is the most convenient way of calculating the results that the demonstration provides.

but whereas the laws of planetary motion seem abstract, the idea of a planet held in its course by a powerful attraction is a graspable image, even if the gravitational force itself remains a complete mystery.

The calculus, as developed by Newton, could be used and was used by him for the solving of a great variety of mechanical and hydrodynamic problems. It immediately became the mathematical instrument for all understanding of variables and motion, and hence of all mechanical engineering, and remained almost the exclusive one until well into the present century. In a very real sense it was as much an instrument of the new science as the telescope.

THE 'PRINCIPIA'

It must have required all Halley's persuasiveness to make Newton, in the two years 1685–6, embody his solution of planetary motions in his

De Philosophiae Naturalis Principia Mathematica. It was printed for the Royal Society and bears the imprint of its President, who was rather surprisingly Samuel Pepys, but the Society was short of money and Halley had to pay for its production out of his own pocket. This book, in sustained development of physical argument, is unequalled in the whole history of science. Mathematically it can only be compared to Euclid's *Elements*; in its physical insight and its effect on ideas only to Darwin's *Origin of Species*. It immediately became the bible of the new science, not so much as a revered source of doctrine – though there was some danger of this, especially in England – but of further extensions of the methods there exemplified.

Newton, in his *Principia*, did far more than establish the laws of motion of the planets. His grand object was certainly to demonstrate how universal gravity could maintain the system of the world. But he wished to do this not in the old philosophical way but in the new, quantitative, physical way. In this he had two other tasks to fulfil: first of all to demolish previous philosophic conceptions, old and new; and secondly to establish his own as not only the correct but also the most accurate way of accounting for the phenomena.

A great deal of the *Principia* is taken up with a careful and quantitative refutation of the system most in vogue and with which he himself had

144. 'Jet engine' driven by a jet of steam, and illustrating Newton's third law of motion concerning action and reaction of bodies. From *Mathematical Elements of Natural Philosophy, confirm'd by Experiments*, London, 1747. This was a translation by John Theophilus Desaguliers of Willem Jacob 'sGravesande's *Physices Elementa Mathematica*. Newton's friend 'sGravesande was professor of astronomy and mathematics at Leyden, where he introduced Newton's new ideas, and his book was a reasoned collection of demonstrations to prove their validity.

145. A photograph of Halley's comet taken at its return close to the Sun and Earth in 1910. The bright star on the right is the planet Venus, much over-exposed. The comet is named after Halley because he computed its periodic future appearances. His computations were based on Newton's gravitational theory and were the first to apply it to a specific problem. Halley had good reason to be familiar with Newton's work since it was he who edited and piloted the first edition of the *Principia* through the press and, indeed, paid for its printing.

flirted, that of Descartes with its set of whirlpools in which each planet was held. This was a genial intuitive idea but one totally incapable, as Newton showed, of giving accurate quantitative results. In doing so, he was led into founding the science of *hydrodynamics*, discussing and refining the ideas of *viscosity* and the resistance of the air, and indeed laying the basis for a mechanics of fluids that was to come into its own only in the day of the aeroplane.

Though Newton used the calculus in arriving at his results, he was careful in the *Principia* to recast all the work in the form of classical Greek geometry understandable by other mathematicians and astronomers. The immediate practical consequence of its publication was to provide a system of calculation enabling the positions of the moon and planets to be determined far more accurately, on the basis of a minimum of observations, than his predecessors could by their empirical extension of long series. Three observations, for instance, were sufficient to fix the position of a celestial object for an indefinite future.

The proof of this was furnished soon after Newton's time by his friend Halley in his famous comet, whose return he successfully predicted on the basis of Newton's theories. As a result of using Newtonian theories nautical tables became far more accurate. Unfortunately, the most suitable celestial object to observe for the purpose of finding the longitude is the moon, and the moon's motion is quite the most complicated in the solar system. It was never reduced to good enough order to be a reliable guide to sailors, and in the end it was the scientifically minded clockmakers who took the prize – or as much of it as they could persuade the Admiralty to part with – from the mechanically minded astronomers.

NEWTON REPLACES ARISTOTLE:
AN ESTABLISHED UNIVERSE AGAINST A MAINTAINED ONE

Newton's theory of gravitation and his contribution to astronomy mark the final stage of the transformation of the Aristotelian world-picture begun by Copernicus. For a vision of spheres, operated by a first mover or by angels on God's order, Newton had effectively substituted that of a mechanism operating according to a simple natural law, requiring no continuous application of force, and only needing divine intervention to create it and set it in motion.

Newton himself was not quite sure about this, and left a loophole for divine intervention to maintain the stability of the system. But this loophole was closed by Laplace (p. 664) and God's intervention dispensed with. Newton's solution, which contains all the quantities necessary for the practical prediction of the positions of the moon and the planets, stops short of any fundamental questioning of the existence of a divine plan. Indeed Newton felt he had revealed this plan and wished to ask no further questions.

He got over the awkward assumption he had made on the existence of absolute motion by saying, following his Platonist friends, that space was the sensorium – awareness or brain – of God, and must therefore be absolute. In this way he avoided confusing himself in relativistic theories. His own theory gave no reasons why the planets should all be more or less in a plane and all go round the same way – for which Descartes' whirlpool had given a facile explanation. Newton honestly disguised his ignorance of origins by postulating that this was the will of God at the beginning of creation.

By this time the destructive phase of the Renaissance and Reformation was over; a new compromise between religion and science was needed just as much as those between monarchy and republic and between the upper bourgeoisie and the nobility. Newton's system of the universe did represent a considerable concession on the part of religious orthodoxy, for by it the hand of God could no longer be clearly seen in every celestial or terrestrial event but only in the general creation and organization of the whole. God had, in fact, like his anointed ones on earth, become a constitutional monarch. On their side the scientists undertook not to trespass into the proper field of religion – the world of man's life with its aspirations and responsibilities. This compromise, wisely advocated by Bishop Sprat, and preached by the redoubtable Dr Bentley in his Boyle sermons of 1692, was to last until Darwin upset it in the nineteenth century.*

Although the system of universal gravitation appeared to be at the time, and still remains, Newton's greatest work, his influence on science and outside it was even more effective through the methods he employed in achieving his results. His calculus provided a universal way of passing from the changes of quantities to the quantities themselves, and vice versa. He provided the mathematical key adequate for the solution of physical problems for another 200 years. By setting out his laws of motion, which linked force not with motion itself but with change of motion, he broke definitely with the old commonsense view that force was needed to maintain motion, and relegated the friction, which makes this necessary in all practical mechanisms, to a secondary role which it was the object of the good engineer to abolish. In one word Newton established, once and for all, the *dynamic* view of the universe instead of the *static* one that had satisfied the Ancients. This transformation, combined with his atomism, showed that Newton was in unconscious harmony with the economic and social world of his time, in which individual enterprise, where each man paid his way, was replacing the fixed hierarchical order of the late classical and feudal period where each man knew his place.*

Quite apart from these actual achievements, Newton's work, itself that final refinement of a century of experiment and calculation, provided a reliable method which could be used confidently by the scientists of later times. At the same time it reassured scientists and non-scientists alike that the universe was regulated by simple mathematical laws. Thus the laws of electricity and magnetism, as we shall see (p. 603), were built on a Newtonian model, and the atomic theory of the chemists was a direct outcome of Newton's atomic speculations.

THE PRESTIGE AND INFLUENCE OF NEWTON

The very successes of Newton carried with them corresponding disadvantages. His abilities were so great, his system so apparently perfect, that they positively discouraged scientific advance for the next century, or allowed it only in regions he had not touched. In British mathematics this restriction was to remain until the mid nineteenth century. Newton's influence lasted even longer than his system, and the whole tone he gave to science came to be taken so much for granted that the severe limitations it implied, which were largely derived from his theological preconceptions, were not recognized till the time of Einstein and are not fully even now.

Paradoxically, for all his desire to limit philosophy to its mathematical expression, the most immediate effect of Newton's ideas was in the economic and political field. As they passed through the medium of the philosophy of his friend Locke and his successor Hume, they were to create the general scepticism of authority and belief in *laisser-faire* that were to lower the prestige of religion and respect for a divinely constituted order of society. Directly through Voltaire, who first introduced his work to the French, they were to contribute to the 'Enlightenment' and thus to the ideas of the French Revolution. To this day they remain the philosophical basis of bourgeois liberalism.

7.10 Retrospect: Capitalism and the Birth of Modern Science

Looking back over the epic movement of the new science in the fifteenth, sixteenth, and seventeenth centuries we are now better placed to see why the birth of science occurred when and where it did. We can see how it followed closely on the great revival of trade and industry that marked the rise of the bourgeoisie in the fifteenth and sixteenth centuries and its political triumph in England and Holland in the seventeenth. The birth of science follows closely after that of capitalism. The same spirit that broke the fixed forms of feudalism and the Church had also broken with the even older slave-owning, conservative tradition of the classical world. In science, as in politics, a break with tradition meant a liberation of human ingenuity into hitherto closed fields. No part of the universe was too distant, no trade too humble, for the interest of the new scientists.

THE UNITY OF SEVENTEENTH-CENTURY SCIENCE

Yet despite the variety of fields of study, science in the seventeenth century had an underlying unity which had a threefold basis: that of persons, of ideas, and of applications. In the first place, the scientist of the seventeenth century was himself able to cover and to produce original work over all the field of then known science. Newton was not only a mathematician, astronomer, optician, and mechanic, but he worked for years on chemistry, of which, though he published little, it appears he had a far deeper understanding than any other man of his time. Hooke, though no great mathematician, worked, as we have seen, in all these fields as well as in physiology, and is one of the pioneers of microscopy. Wren, whom we know as an architect, was also at the very centre of the scientific movement. As a result of this universality the scientists or *virtuosi* of the seventeenth century could get a more unitary picture of the field of science than it has been possible to achieve in later times.[4.95]

MATHEMATICAL PHILOSOPHY

In the second place, there was an underlying unity produced by a guiding idea and method of work that was essentially mathematical and based on a mathematics derived directly from the Greeks, but including also Arabic, Hindu, and possibly Chinese contributions. This was not sheer gain; an effective, though unrecognized, limitation of the field of seventeenth-century science was due to this preoccupation with mathematics. Those parts of experience that could not then be reduced to mathematics tended to be left out, and even those parts which were not suitable for mathematics tended to be treated mathematically, with somewhat ridiculous results. A follower of Harvey, for example, tried to explain the action of the different glands of the body by the relative momentum of their particles, which depended on the angles at which their ducts discharged. The extreme case was in the social field, with the attempt by Spinoza (1632–77), the noblest of the seventeenth-century philosophers, to reduce ethics to mathematical principles. It was because of the insistence on mathematics that the scientists of the seventeenth century succeeded only in those fields, such as mechanics and astronomy, where the Greeks had been before them, and made little significant progress in chemistry and biology.

SCIENCE AND TECHNICAL PROBLEMS

The third and most characteristic unifying principle of the new science was its concern with the major technical problems of the day. As we have seen, the enormous advance of technique from the fourteenth century, or even before, arose out of the break with tradition in the favourable circumstances of Europe, where abundant resources had to be exploited by few men, thus putting a premium on ingenuity. The solutions reached in mining and metal-working, transport and textiles, were technical solutions, but by breaking with tradition they raised new problems which modern science was created to solve. Enough of these problems, especially those of navigation, gunnery, and mechanics, lay within the scope of the Greek tradition of learning to be within range of immediate practical solution. The remainder were to form the inspiration of eighteenth-century science.

SCIENCE PROVES ITS WORTH

It is true that at first the scientists claimed to be able to achieve greater results than were possible at the time. Until the end of the eighteenth century science drew far more from industry than it could yet give back. In chemistry and biology it was to be at least another hundred years before anything that the scientists could propose could replace or improve on the traditional processes, in medicine even longer. Even among the well-understood physical sciences, both in mechanics and gunnery, the practical man still held the advantage. The improvement of mills was for long to be in the hands of the millwrights, that of guns in those of the founders. Working in wood or in roughly cast metal it was impossible to make use of the refinements which the new mathematics and dynamics could provide. Newton, for instance, did work out the trajectory of shot allowing for the resistance of the air. His methods were still being used in the Second World War, but they were quite inapplicable in his time. Gun barrels were uneven, the shot did not fit them, the quality and quantity of powder varied with every filling, and there was no means other than rough manhandling with ropes and wedges for pointing the gun. The

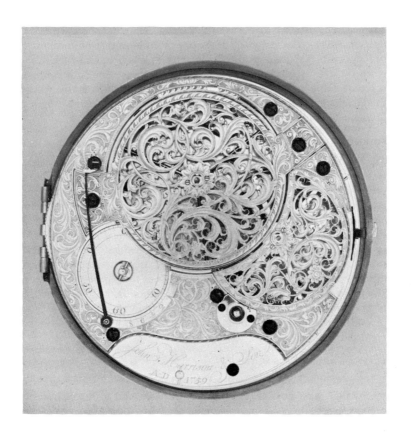

146a, b. Harrison's marine chronometers, numbers 1 and 4. John Harrison (1693–1776), trained originally as a carpenter, solved the problem of determining longitude at sea by designing and building the first really accurate timepieces that could be successfully used on board ship. His first chronometer was a large clock with two double pendulums, each of which had a bob at either end of its arms. It also possessed a temperature compensating device, vital for a clock that was to keep accurate time, but then something of an innovation. Harrison improved his chronometer over the years, and his fourth design was not much larger than a pocket watch, yet it kept time to within one-tenth of a second per day, and in 1764, contrary to general opinion at the time, proved longitude determination using a timepiece to be eminently practicable.

practical gunner, who knew the limitations of his art, could well dispense with ballistics.[4.82] The only exception to this was the art of the clock-maker, in the higher reaches of which – the design of marine chrono-meters – some knowledge of dynamics was a necessity.

The one great success of the new science lay in navigation. This was achievement enough, for it was at a time when control of the sea-ways and the opening up of the new world were the key to national, economic, and political success. By proving its worth there, science became an established part of the new dominant capitalist civilization. It acquired

147. In spite of the developments in theoretical mechanics of Newton and others, the practical engineer had still to rely on human or animal power, or on wind and water, as a driving force. From George Andreas Böckler, *Theatrum Machinarum Novum*, Nuremberg, 1673.

a continuity and a status that it was never to lose. The importance of science was to grow relatively and absolutely as it came to be realized that the military and economic superiority of European civilization over the old civilizations of Islam, India, and China, was due to its technical achievements, and that the improvement of technique required the continuous application and development of science.

ANCIENTS AND MODERNS

It was in this field of techniques that the men of the seventeenth century felt superior not only to their forebears of the Renaissance and of the barbarous Middle Ages, but even to the almost legendary achievements of the ancient Greeks and Romans. Modern men, it was felt, might not be wiser or better, but they were certainly more ingenious and could do things the Ancients never dreamed of, like shooting off guns or sailing to America. More important than the achievement itself was the knowledge that it was only a beginning, that there was no limit to possible advance along the same line. As early as 1619 Johan Valentin Andrae, Comenius' tutor, had declared, 'It is inglorious to despair of Progress', and that idea, so foreign to the medieval, if not entirely to the classical mind, was launched on its triumphant career.[4.77]

Indeed it was towards the end of this period that the battle between Ancients and Moderns was most consciously engaged. It ranged with varying fortunes all over the world of knowledge.[4.95] Its most famous expression was Swift's *Battle of the Books* where the Moderns certainly got the worst of it. But Swift here, as in *Gulliver's Travels*, was swimming against the stream. However much they might still ornament the libraries of gentlemen the classics were for all practical purposes dead. They might still be authorities for the composition of sonorous prose but they had nothing to contribute to philosophy as the eighteenth century understood it.

Progress was still rather an ideal than an achievement. The great transition of the fifteenth, sixteenth, and seventeenth centuries had not brought about any revolutionary change in the material mode of life. That was still to come. Wealth and poverty had been redistributed. There were far more well-off people in England and the Low Countries at the end of the period than at the beginning, though probably fewer in Italy. What was important was that the method of multiplying wealth by turning it into capital had now broken through the feudal restrictions and the way was open to its indefinite extension. Under capitalism in its first phase the new incentive of profit was putting a premium on technical advance. The financial structure was, however, top heavy and unstable from the start. The merchants and gentlemen of the seventeenth century, for all their wealth and occasional interest in science, were not the men to make use of the new possibilities; but they had cleared the ground for the flourishing of a humbler set of manufacturers, who were, thanks to science, to make use of and develop the traditional techniques of civilization out of all recognition.

THE INTELLECTUAL REVOLUTION

It would, however, be entirely wrong to consider the driving force of science as completely utilitarian. Science still carried much of the prestige, political and ethical, of the philosophy of the ancient world to which the Renaissance had added so greatly. Natural philosophy, as it was called, was a worthy, even noble profession and its patrons, in supporting it, were adding lustre to the State. The men of the new experimental science felt that it was they, rather than the schoolmen, who were the true heirs of the Ancients; and in fact the only parts of the external world where their methods succeeded were those already cultivated by the Greeks. Nevertheless, while the mathematics of Greece was one characteristic tool of modern scientific method, the whole intellectual movement of science arose out of the struggle against the philosophy of Greece, adapted as it had been in the Middle Ages to the service of a now outmoded feudal system. In its early phases the new experimental science was necessarily critical and destructive; in its later phases it aimed at providing a new basis for a philosophy more in tune with the needs of the times. The break was never complete; the hold of religion, both internal and imposed by society, was still too great to allow much deviation from the general scheme of creation and salvation accepted by Catholics and Protestants alike. Explicitly with Bacon and Descartes and even in the more cautious implicit philosophy of Galileo and Newton great liberties were nevertheless taken with the scheme of the divine governance of the world. These were in the next century to be the basis of criticism of the whole framework of religion.

The paradox of the Scientific Revolution was that those who contributed most to it, substantially the scientific innovators from Copernicus to Newton, were the most conservative in their religious and philosophic outlook. If they were not orthodox it was only because they thought that orthodoxy had wandered from the path of reason. They accepted the programme of St Thomas Aquinas of reconciling faith and reason, but they were forced to reject his conclusions because the scheme of the world to which he had reconciled his faith was now revealed to be palpably absurd. Their own forms of reconciliation were to prove even less durable. But the day of theological domination over science had ended. It could still distort and delay the advance of science, but it could not stop it. Religion was tacitly confined to the moral and spiritual sphere. In that of the material world the Scientific Revolution, willed or unwilled, had definitely taken place.

SCIENCE ESTABLISHED

By 1690 science had definitely arrived. It had acquired an enormous prestige, at least among the upper ranks of the society of the time. It had its organization in the Royal Society and the Académie Royale des Sciences, which were closely linked by personal ties with the ruling powers – with Parliament and the great Whig houses in England, with the Royal Court in France. It was spreading to other countries. A coherent discipline of experiment and calculation had been evolved, a coherent method by which any problem could sooner or later be tackled. The foundations of science might later be underpinned and altered, but the edifice raised on them was stable, and, even more important, the general method for raising it was now known and was never likely to be forgotten again.

However, the very success of the early scientific method had elements of danger in it. The method itself incorporated much of older ideas which inevitably coloured the thoughts of the first scientists, and enshrined them, with the new conceptions derived from experiment, in the new philosophy of science. It is this unconscious relic of the past that is now appearing in much of the idealistic scientific theory of today; and it may well be that the task of twentieth-century science will be to break up the system of Newton just as the seventeenth century broke up that of Aristotle.

Table 4

The Scientific Revolution (Chapter 7)

This table (next page) attempts to present some of the major features of the birth of modern science in their relation to political, economic, and technical developments. The time-scale of the period, 1400–1700, is uniform, but the phases corresponding to the sections of Chapter 7 are indicated. This brings out the great concentration of effort in the last of these phases. The major critical discoveries and theories, such as Copernicus's vindication of the solar system, Harvey's circulation of the blood, and Newton's theory of gravitation, are specially indicated. The table has been drawn up to bring out the most significant relationships. Owing to their complexity, however, other relations, such as those between Harvey's discovery and the study of pumps, are not shown here, though some of these are given in Table 8 (Volume 4).

	Historical Events	Philosophy	Navigation	Mathematics and Astronomy
	Authorities Reinstated→PLATO		GERSON	ARCHIMEDES ARISTARCHUS
	Dethroned→ARISTOTLE			PTOLEMY
1440	Italian Renaissance Platonic academies in Florence Great growth of trade and arts	Humanism return to the classics	Portolan maps School of Sagres Portuguese along African coast	Recovery of Greek mathematics *Peurbach* revival of astronomy
				Müller
	Italian wars		Columbus discovers America *Vasco da Gama* reaches India	nautical almanacs
1500	*Francis I* College de France Reformation *Luther* *Calvin*	*More* 'Utopia' *Vives, Erasmus, Rabelais* criticism of medievalism	Magellan round the world	
1540				**Copernicus** SOLAR SYSTEM
1550	Great inflation		*Nunez* maps and navigation	*Tartaglia, Cardan,* algebra revived
	Counter Reformation Religious wars in France		Problem of longitude	
	Revolt of Netherlands	*Montaigne* scepticism	*Mercator's* maps	
	Elizabethan age Gresham College	*Bruno* plurality of worlds	*Norman* magnetic dip	*Vieta* symbolic algebra *Tycho* accurate observations
1600	Capitalism coming to power Accademia de Lincei Thirty Years War Civil wars in Britain Informal meetings of scientists	**Bacon experimental philosophy** **Gassendi atomism** **Descartes mechanical philosophy** *Hobbes* materialism	**Gilbert on magnet**	**Kepler planetary orbits** *Napier* logarithms *Descartes* analytical geometry *Fermat* number theorem
1650	Royal Society Louis XIV in France	*Spinoza* rational morality	*Guericke* frictional electricity	**Newton calculates** THEORY OF GRAVITATION
	Académie des Sciences			*Leibniz* differentials
	Expulsion of the Huguenots			
1690		*Leibniz* pre-established harmony		
			↓ Electricity	↓ Mathematical Physics

Chapter 7.1–7.3

Chapter 7.4–7.6

Chapter 7.7–7.9

Optics	Mechanics and Hydraulics	Chemistry	Medicine, Physiology, and Natural History
ALHAZEN	PHILOPONOS	LULL	ARISTOTLE
	ARISTOTLE		GALEN
Developments in painting and perspective	Developments in metallurgy, mining and pumping	Beginnings of chemical manufacture, alcohol, gunpowder, alum	
		Alchemy turns to chemistry	
	Leonardo da Vinci		**Leonardo da Vinci**
Scientific painting	Engineering, water-works		Drawings of anatomy and natural history
Dürer perspective	Development of gunnery	**Paracelsus revival of chemistry** *Agricola* 'De Re Metallica'	*Paré* surgery **Vesalius 'De Fabrica'**
	Tartaglia ballistics		*Servelus* pulmonary circulation Collection of rarities Development of gardening and agriculture
Spectacle-makers invent telescope	Development of dykes, canals, and locks, in Holland *Stevin* statics and hydraulics		
Galileo Telescopic observation, 'Two chief systems' Trial	**Pendulum 'Two new sciences', Dynamics**	*Van Helmont* gas	
	Scientific study of pumps		**Harvey circulation of blood** generation of animals *Leeuwenhoek* microbiology *Malpighi* microscopical anatomy *Mayow* theory of respiration *John Ray Nehemiah Grew* Classification of animals and plants
Newton theory of colour *Römer* velocity of light *Huygens* wave theory of light	**Torricelli barometer Guericke vacuum Boyle gas law** *Hooke* experimental physics	*Boyle* 'Skeptical Chymist' Combustion	
↓ Optical Instruments	↓ Steam Engine	↓ Rational Chemistry	↓ Scientific Biology

PART 5

Science and Industry

Introduction to Part 5

The eighteenth and nineteenth centuries were the great formative centuries of the modern world, centuries that appeared to those who lived in them as representing a liberating phase of human development in which man had at last found the true way to prosperity and unlimited progress. To us, with the experience of the disturbances and changes of the twentieth century, they appear as centuries of preparation, centuries in which great things were done at the expense of much human suffering to produce a grandiose but unstable culture. They cover the period of the establishment of science as an indispensable feature of a new industrial civilization. The new methods of experimental science elaborated in the seventeenth-century revolution were to be extended over the whole range of human experience and at the same time their applications were to keep pace with and infuse the great transformation of the means of production which we call the *Industrial Revolution*.

The Industrial Revolution was not mainly, and certainly not in its first phases, a product of scientific advance, though certain contributions of science, notably the steam-engine, were to be essential ingredients in its success. Nevertheless the whole movement was far more closely identified with the growth and inner transformation of the economic system of *capitalism*, from the phase dominated by merchants and small manufacturers to one dominated by financiers and heavy industry.

It is no accident that the intellectual formulations of science, the technical changes of industry, and the economic and political domination of capitalism should grow and flourish together at the same times and in the same places. The relations between them are, however, by no means easy to unravel. Techniques, economic forms, and scientific knowledge were all growing and changing rapidly in the period; sometimes one seems to take the lead, sometimes another. It will be our task in this section in particular to try to disentangle the contributions of science to technical and economic transformations as well as to trace the effect of these transformations on the growth and character of science itself. This, however, can become apparent only after a more detailed study of particular aspects of interrelation, and conclusions can be discussed only at the end of the section.

At the outset, however, it is necessary to give a broad description of the social and economic changes of the period so that those of science can be seen in adequate perspective. Already by the end of the seventeenth century the stage was set for the further advance of the new – capitalist – mode of production. In what was still but a small corner of Europe, almost limited to England, the Low Countries, and northern France, the urban middle class had broken away to a greater or lesser degree from feudal limitations; they could finance production for profit with an ever-increasing market for their products all over the world which the new navigation had opened to them. Production was still handicraft and domestic, but merchants and capitalist *manufacturers* were coming to control it more and more, and both craftsmen and peasants were being depressed to the status of wage labourers.

With the combination of an expanding market, growing freedom from manufacturing restrictions, due to a break-up of urban guilds, and a field of investment in profitable enterprise, there was a premium on technical innovations such as textile machinery, and also on revolutionary scientific inventions such as the steam-engine, which could cut costs and expand production and profits. Better organization of labour, the division and specialization of tasks, the factory system, and ultimately power-driven machinery, were all means to this end, and all drew from it the social drive necessary to break down the older-established systems of production. Once this process started in the latter part of the eighteenth century it tended to grow and spread to other fields by its own success based firmly on the new capital it generated. By the mid nineteenth century the domination of capitalism over the whole world was unquestioned, but that very fact did more than put a limit to its expansion. It made evident a fundamental instability from which it could not escape. By its very nature production for profit could never allow a sufficient share of goods or opportunities for the vast new population of wage labourers that it had brought into existence to provide for a continuous prosperity (pp. 1134 f.). Booms were followed by slumps of increasing severity and the competition for limited markets provoked international rivalries. The open breakdown of the system was, however, not to begin till the twentieth century. For most of the period we are discussing here, the progress of science occurred against a background of an expanding industrial capitalism which tended to make more and more calls on it.

TECHNIQUE AND SCIENCE

Though in their first stages changes of technique in response to economic needs could and did take place without any intervention of science, it often happened that the mere following of existing trends led to unforeseen difficulties which could be removed only by invoking science. For example, a natural source of supply such as a vegetable dye might run short, due merely to an increased production of cloth, thus creating a demand for an artificial substitute that could be found only through the help of science (pp. 631 f.). Or to take another example, the transition from home to large-scale brewing might in itself provoke disastrous failures which could be prevented by an appeal to science (p. 647).

This ancillary, almost medical, role of science in industry was, towards the end of the nineteenth century, replaced by a more positive one. Ideas originating in the body of science itself were developed to form new industries. The first and most important of these was the steam-engine – the *philosophical engine* of the early eighteenth century; but once its general principles became familiar its manufacture and use were absorbed into practical engineering. It is only at the end of the nineteenth century that industries that started and remained scientific, such as the chemical and electrical industries, began to take form, and their full development was not seen till the twentieth century.

Despite the contribution of the steam-engine it cannot be claimed that science was a major factor in effecting the decisive change from hand to machine production that took place in the last quarter of the eighteenth century. This new method of production proved to be, on the other hand, a great forcing house for scientific knowledge. In the nineteenth century the situation began to alter. Science came to be a major agent for effecting technical developments. Its full integration into the productive mechanism had to wait till the twentieth century.

The relation of science to the history of the period is, however, by no means confined to its role in the productive process. The new form of society based on money exchange was taking form, with its emphasis on liberty and individual enterprise in contrast to the fixed status and social responsibility of the Middle Ages. This society, limited by class and by country as its benefits were, required a new set of ideas to express and justify itself. It found them, to a large extent, in the methods and results of the new sciences, while they in their turn were profoundly, though unconsciously, influenced in the formulation of their theories by prevailing social beliefs.

THE SCIENTIFIC AND INDUSTRIAL REVOLUTIONS

It may appear somewhat arbitrary to divide, at the beginning of the eighteenth century, an Industrial Revolution from the Scientific Revolution of the seventeenth. There is naturally no question of the unbroken continuity between them. It might seem better to treat them as successive phases in one great transformation. Nevertheless, it seems to me that the distinction is more than one of convenience. There is a noticeable difference in quality between the two periods. The break-through in the former was essentially in understanding, in the second in practice. It is tempting to think of this as a relation of cause and effect, but the real relations are, as I hope to show, far more complex. To a certain extent the two evolutions of knowledge and power ran in parallel, driven by separate internal influences, though always reacting on each other, especially in periods of rapid advance (pp. 1221 f.). Towards the end of the seventeenth century a third, economic factor, the expression of capitalism in manufacture, makes itself felt. It is to this that we may look for the transition of the mathematical–astronomical–medical science of the seventeenth century to the chemical, thermal, and electrical science of the eighteenth. The nature of the complex interaction of science, industry, and society will, I hope, be more evident from actual examples of the history of this interaction contained in the following two chapters.

PHASES AND ASPECTS OF
THE GROWTH OF INDUSTRY AND SCIENCE

In order to follow these interactions concretely over a period so rich and complex, without losing sight of the unity and continuity of the historic process, the best method I have been able to find is to operate here a double system of division by period and by subject, providing a kind of cross classification. These two divisions will be found successively in Chapters 8 and 9, followed by a general conclusion.

The division by sub-periods is, in Chapter 8, a particularly difficult one, partly because the wealth of information available prompts minute subdivisions, but even more because of the impossibility of finding divisions applicable at the same time to political, economic, technical, and scientific history. Politically, for instance, the great divide is most evidently that of the French Revolution and the Napoleonic wars; these, however, provoked no loss of continuity but rather a general enhancement of scientific activity. The decade 1760–70 is, on the other hand, a turning point in technical and scientific history (pp. 549 f.), but is not so noticeable in the political sphere. Sometimes the divisions nearly coincide, as in 1831, when reforms in politics and science come together – by no means accidentally, as they were advocated by the same men and supported by the same popular movements (pp. 549 f.).

My final choice has been to divide the whole period into four major phases. First the transitional or latent phase (8.1) leading up to the Industrial Revolution, that is 1690–1760. The second phase (8.2–8.4) includes the whole of the French Revolution, 1760 to 1830. This phase is as revolutionary in technique and in science as in politics, covering as it does the major advances of the Industrial Revolution and the Pneumatic or Chemical Revolution, second only in importance to the Mathematical–Mechanical Revolution of the seventeenth century.

The third phase (8.5–8.6) is the mid nineteenth century, from 1830 to 1870, what has been called the heyday of capitalism. Finally the fourth phase (8.7–8.8) is a very short one, from 1870 to 1895, which in the outer world marked the beginning of modern imperialism and in science the transitional period before the great twentieth-century revolution.

The second and third of these phases include two notable periods of advance and triumph of science. The first was, after the heroic age of the seventeenth century, rather a scientific backwater, a taking breath and preparing for the advance that was to come. And, in a different way, so was the fourth phase, though in both cases those working at the time felt they were completing a grand edifice: in one case the edifice of Newtonian physics, in the other the great physical synthesis of Faraday and Maxwell, and the great biological syntheses of Darwin and Pasteur.

Even with such a division of the period the general surveys of science in its historic setting, such as have been given in earlier chapters, will no longer suffice to provide an adequate picture of its now increasingly separate disciplines. For this purpose an attempt is made in the succeeding chapter (9) to follow out the development of five of the major lines of technical and scientific advance over the whole period of the eighteenth and nineteenth centuries. Those selected are: 9.1, Heat and Energy, including the history of the steam-engine; 9.2, Engineering and Metallurgy, with particular reference to iron and steel; 9.3, Electricity and Magnetism; 9.4, Chemistry; and 9.5, Biology.

In each section the aim is to bring out the inner coherence and continuous tradition of the field of activity, to illustrate the interplay of economic, technical, and scientific factors, and to bring out the interrelations of different sciences and techniques. Only after both time and subject divisions have been completed will an attempt be made to combine the two approaches and to try to draw from them general conclusions about the position and influence of science in this decisive period of social and scientific transformation.

Chapter 8

Antecedents and Consequences
of the Industrial Revolution

8.1 The Early Eighteenth-Century Pause 1690–1760

The original impetus that had created science in the Renaissance and carried it through the great outburst of the mid seventeenth century seemed to falter and die away towards its end. Within a few years of the publication of Newton's *Principia* in 1687, indeed even before it was written, there was a perceptible slackening of scientific effort and dying away of curiosity. This dip in the curve of scientific progress was a general phenomenon and not merely confined to England, though naturally, because science had been so highly developed there in the early days of the Royal Society, it was most clearly to be seen there.

To some extent this pause might be ascribed to reasons internal to the scientific world. The prestige of Newton had turned science in a direction that was to be sterile for many years because of the very finished character of Newton's own work and the distance by which he surpassed his contemporaries. To a far larger extent, however, the slackening of scientific advance in England, and to a lesser degree in the rest of the learned world, was due to social and economic factors. The class that had started the seventeenth-century scientific drive, the gentlemen merchants who were then concerned with using new methods based on science in navigation, trade, and manufacture, had been succeeded by a new generation, wealthier, less enterprising and curious, and much more complacent. These, the first Whig aristocracy, found the most secure investment in land and an outlet for their speculative interest in such glorious gambles as the South Sea Bubble.[4.46] The class that was to replace them in power, the rising but still small manufacturers who were later to create the Industrial Revolution, had not yet become conscious of the possibilities or even of the existence of science. They were occupying themselves throughout the early part of the eighteenth century with developing and using improved technical methods, still for the most part hand operated, which served for a while to cope with the ever-increasing demands for cloth and manufactured articles.

148. The collapse of the gamble of the South Sea Bubble brought much suffering and came in for much adverse comment. It led William Hogarth to prepare one of his sarcastic symbolic engravings.

These changes reflected themselves in the Olympus of science, the Royal Society; the impetus to serve trade withered away and the Society itself fell on some very lean days. Conrad von Uffenbach, who visited the Royal Society at Gresham College in 1710, writes of its collection of instruments as

not only in no sort of order or tidiness but covered with dust, filth and coal smoke, and many of them broken and utterly ruined.

He continued:

If one inquires after anything the operator who shows strangers around will usually say 'A rogue had it stolen away', or he will show you pieces of it, saying 'it is corrupted or broken'; and such is the care they take of things.[5.15]

The society was in serious financial difficulties and an inquiry of 1740 showed that a large number of fellows had ceased to pay their subscriptions.[4.19]

Meanwhile, however, though science somewhat languished, technical change had not ceased, and if the advance in the early eighteenth century seems slow it is only in relation to the vast changes that were effected in a few decades by the Industrial Revolution. Some of these lines of change which were well under way in Britain during the early part of the century were to be of the utmost importance for the future both of industry and science.

One of these was the rapid improvement in agricultural practices. These improvements, adapted from those of the Dutch in the seventeenth century (p. 474), spread rapidly in Britain, and helped to make commercial farming pay. They were made possible on the one hand by the availability of capital, originally from mercantile sources, to invest in land, and on the other by the rapid growth of towns, in the first place of London, that provided a reliable market for corn, meat, and vegetables. Technically an advance, they were socially unjust and cruel, involving the ejection by Enclosure Acts of a peasantry with traditional but poor title to the land and with even poorer means of cultivating it.[5.57]

Another change of vital importance was the rapid expansion of a new heavy industry, based on coal, with improved mining and transportation

149. The shift to a coal-based economy altered the balance between northern and southern England which was hastened by the advent of the steam engine. This lithograph of a Yorkshire miner comes from George Walker's *The Costume of Yorkshire*, a collection of prints prepared *c.* 1814 and reprinted at Leeds in 1885. Pit-head machinery is shown in the background and the locomotive is that of John Blenkinsop, the first steam engine to be used for hauling loaded waggons. Blenkinsop's engine came into use around Leeds in June 1812.

methods, and of radically new methods of making iron and steel. Here one scientific development, the steam-engine, originally used for draining mines (pp. 576 f.), was of key importance, as was the technical development of making iron with coke from 'pit' coal, instead of using the immemorial wood charcoal (p. 412), which was first effected in an inconspicuous way by the Quaker Abraham Darby in 1709. These developments were, however, limited to what were then minor fields of industry, and did not amount in themselves to an industrial revolution, though they were its necessary precursors.

This phase marks the actual point of no return between the age-old country-based economy to one based on the coalfields; from an economy of food to an economy of power. In Patrick Geddes' terms it was passing from the era of *eotechnics* to that of *paleotechnics*.[5.49] This, however, is true only of the very growing points of the new technique, on and near the coalfields themselves. The radical changes were largely confined to Britain, though in the iron-making countries there was an independent development of machinery, as in the rolling and slitting mills of Polhammer (1661–1751) in Sweden[5.92; 5.12.635] and the use of a steam-engine for iron working by Polzunov (1758) in the Urals.[5.32]

The shift to a coal-based economy was not only to alter the balance between northern and southern England but also to be a major factor in the meteoric rise of Scotland as an industrial and intellectual power of the first rank.[5.4] Scotland, despite the antiquity of its traditions and the Calvinist movement of the sixteenth century, had not kept pace with the rapid development of England in the seventeenth. The resources for the early Industrial Revolution were lacking. The position was very different once the advantages of coal were realized. The very poverty of the country, combined with the high literacy and puritan traditions, meant that once the idea of improvement was accepted it would not be held back, as it had been in England, by complacency and ignorance.

Moreover, also due to Calvinism, Scotland had established an intellectual link with Holland, particularly with the university of Leyden, ensuring a steady flow of well-trained men, especially in medicine, which included chemistry. The great Boerhaave (1668–1738), a follower of van Helmont and teacher of half the chemists of Europe, had a particular influence on Scotland, where his pupils took the leading part in introducing science to the universities. The universities of Scotland, in the eighteenth century, were indeed most unlike their English sisters; they became active centres of scientific advance which sought in every way to link practice to theory (p. 529).[5.66]

FASHIONABLE SCIENCE IN FRANCE: THE PHILOSOPHES

While Scotland and England were rapidly approaching the Industrial
Revolution, the developments of even such an advanced country as
France still continued along the old lines. There was a steady growth of

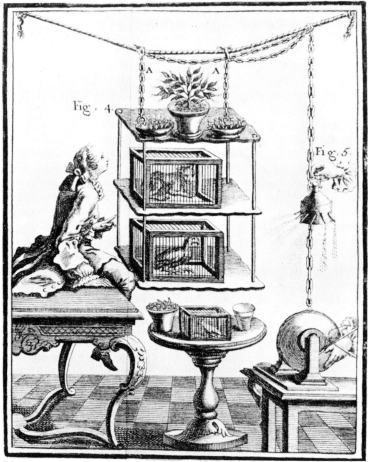

150. In France science, and particularly electrical science, became a fashionable
pastime for the Court. An engraving from the Abbé Nollet (1700–70), *Récherches
sur les Causes Particulières des Phénomènes Électriques*, Paris, 1753. In the lower
right of the picture there is an electrical machine; the dotted lines represent electrical
discharges and the chain carries high voltage electricity to the items suspended from
silk cords in the centre.

handicraft industry of very high quality with considerable division of labour and an output greater than that of England, but there was no attempt to use large-scale machinery except for such purposes as royal waterworks.

Nevertheless, the same period in France saw a sudden rise of activity in science, though this rise was of a very different kind from that in England. It was essentially an expression of interest, on the one hand, of part of the rather bored aristocracy, not, as in England, occupied practically with its estates, but cooped up in court circles; and on the other, a mode of expression of dissatisfaction with affairs on the part of a rising middle class, headed in France by the administrative and legal professions. Science was fashionable and revolutionary at the same time. It is symptomatic that the man who introduced Newtonian philosophy into France was none other than Voltaire (1694–1778).[5.77]

Much of the effort of the amateurs of science – natural philosophers or *philosophes* – was spent on criticism of existing institutions which were felt to be cramping the economic and political development of the country. There was, however, an increasing interest in industry but, unlike in England, it came from above on the seventeenth-century pattern. Réaumur (1683–1757), for example, a man of great intelligence and wide interests, carried out from 1710 to 1720 a long industrial research on steel-making (p. 596). Because, however, they met with no response in a tradition-ridden industry, the results of his discoveries did not create a steel industry in France and their advantages were only reaped by English steel-makers more than a hundred years later.

THE SPREAD OF SCIENCE IN EUROPE: PRUSSIA, SWEDEN, RUSSIA

It was also in this period that interest in science spread far more widely than to the group of countries of France, England, and Holland which had monopolized it in the seventeenth century. Academies on the English and French models were being set up in various kingdoms of Germany and Austria through the efforts of Leibniz, the universal philosopher; and later under the patronage of the eccentric, scientific, and poetical king of Prussia, Frederick the Great. By the middle of the century no court could be called complete without its Academy of Arts and Sciences in which academicians, usually rather irregularly paid, had to compete for princely favour by producing laudatory odes or amusing experiments.*

The northern countries of Sweden and Russia also marked their new military and economic importance by the setting up of academies. These were, however, to have from the outset a different function from the polite societies of the other European countries. They were concerned largely with the scientific study of the great raw material resources of wood, tar, and flax, of iron and other minerals, all so much needed on account of the rapid increase in sea-borne trade that those countries were just beginning to exploit. Peter the Great introduced science as one aspect of his design to create an economically and militarily independent Russia.[5.135] Though at first he had to staff it with foreigners, mostly Germans and French, but including a prince of mathematicians, the Swiss Euler (1707–83), he aimed at building up a truly national body of science. Success was not to come till after his reign with the life-work of that intellectual giant of the eighteenth century, Michael Lomonosov (1711–65), poet, technician, and physicist, the first of a succession of great Russian men of science (p. 620).[5.85; 5.103]

THE ESTABLISHMENT OF SCIENCE:
THE INFLUENCE OF NEWTON

It is not surprising, in view of these social and cultural changes, that the trends in science throughout most of the eighteenth century should be different from those of the seventeenth. In a more gentlemanly age the accent on the useful was not so stressed, though it was never absent, as Reaumur's and Hales' research show (p. 620), and it was to become more prominent than ever towards the end of the century. At the beginning, the entertaining and instructive side of science came more into prominence. There were no more battles with the Churches, which, Protestant and Catholic alike, had lapsed into a tolerant indifference. In any case science had arrived; it was an institution, it had acquired its own internal tradition.

Thanks to Newton, mathematical astronomy was well established as the senior branch of science and it was steadily followed up throughout the century, more successfully in France than in its native England, where the great man's prestige was more paralysing. Nothing, in fact, of physical significance was added to the Newtonian theory, but the mechanical principles were generalized and were combined with a new mathematics due largely to Leibniz. This combination was to prove an instrument for solving the more intricate problems that arose later in the branches of physics, particularly from the study of electricity and of heat. The great generalizations of mechanics of Euler, d'Alembert, Maupertius, Lagrange and Laplace were to be the basis of the mathematical–physical revolution of the twentieth century.

NEW INTERESTS: ELECTRICITY AND BOTANY

Though these studies carried the full prestige of science, the immediately significant advances lay not in deepening but in widening its field of interest. The major contributions which were made to science in the early and middle eighteenth century were in the fields of *electricity* and *botany*, one an entirely new addition to science, the other a revival and a new formulation of almost the oldest of the sciences. Both, in their first stages, showed a definite trend away from the mechanical and mathematical bent of the seventeenth century into fields of greater variety and less rigour (pp. 599, 636).

The study of electricity started as rather a pleasant and useless pastime and provided a series of new, exciting, and spectacular experiments. It was Franklin who, by his invention of the lightning conductor, literally brought electricity down to earth and forecast its future importance. Botany escaped in the eighteenth century from the care of the herb garden from which the doctors of medicine prepared their physic and, under the inspiration of Linnaeus, spread everywhere into the wilderness, reinforcing the social tendencies of a bored aristocracy and a thwarted bourgeoisie to return to Nature.

With botany came a renewed interest in collections of all kinds – coins, minerals, fossils – very suitable for noblemen's cabinets, later to blossom forth as new museums. The curators came to form a new group of scientists, ranging from wealthy and eminent Sir Hans Sloane (1660–1753) whose magnificent collections were the nucleus of the British Museum,[5.20] to the light-fingered Raspe (1737–94), who has the double distinction of being expelled from the Royal Society and writing Baron Munchausen's tales.[5.26]

THE NEW ORDER IN PHILOSOPHY

The early eighteenth century was predominantly a time for the digestion of and reflection on the enormous scientific advance of the seventeenth. The philosophers of the seventeenth century had the task of proving that an alternative existed to the classical–religious world-picture of the Middle Ages, and found it in the prophetic works of Bacon and Descartes, acclaiming the triumph of the new science. Those of the eighteenth century, on the other hand, could take the scientific world-picture which Newton had given them for granted. Their task was to extend it and to reconcile its findings, and still more its attitude of mind, with the new political and economic pattern that was beginning to appear in their own time.

151. Benjamin Franklin's discovery of the value of the lightning conductor led, as so often is the case with new inventions, to all kinds of extravagances. Jacques Barbeu-Dubourg, a member of the Académie des Sciences, even went so far as to invent a portable lightning conductor built into an umbrella. From Louis Figuier, *Les Merveilles de Science*, Paris, *c.* 1870.

At first they preached an attitude of acceptance to a new and rational order. Locke, himself a scientist and doctor, leaving little space for the supernatural, applauded the rule of law – the scientific law of Newton and the civil law established by the constitutional revolution of 1688. Leibniz, for all his mathematical and philosophic gifts and his pleas for European peace, was essentially a medieval thinker. He propounded the doctrine of 'pre-existing order', little different from the Providence of the churchmen, and he applauded the fact that 'everything was for the best in the best of all possible worlds'.[5.86]

Nevertheless, this world would not stay still. The succeeding philosophers felt there was something wrong with this complacent picture. The idealist Irishman, Berkeley, in the interests of established religion, denied the reality of the world and of science except in the eye of God. This produced little effect in his time but was to become a basis of reaction in the twentieth century. The sceptic Hume was much more successful in proving that we could know nothing with certainty, including particularly the dogmas of religion. The cynic Voltaire went further and led the attack on the Church itself in the name of reason and benevolence. As the century wore on philosophy was tending to occupy itself more and more with social and economic reform and to pave the way for the French Revolution (pp. 1054 f.).

8.2 Science and the Revolutions 1760–1830

The second phase of our period covers seventy years as decisive in science as they were in politics. Comparable to the seventeenth century in its scientific importance, it far exceeded it in its immediate and in its practical consequences. It includes the Industrial Revolution in Britain and the political revolutions in America and France. The revolutionary wars in effect divided it in two, though they do not cut the continuity of science and technique. The first forty years, 1760–1800, witnessed all these events and also the onset and consummation of another revolution in science, the pneumatic revolution, which, linked with the discovery of the production of the electric current, was virtually to create a new and rational chemistry. The second part of the phase, from 1800 to 1830, though not so fruitful in new scientific or political ideas, remained one of immense vigour and expansion in all fields of practical human activity.

The connexion between these different aspects of social change cannot

152. The Great Hall of Marshall's flax mill at Leeds, *c.* 1870. Although a 'late' picture, this illustrates well the separate pieces of belt-driven machinery with the operatives at work under their overseer. From *Great Industries of Great Britain*, London, undated.

have been a chance one. Indeed, the more closely they are examined the more intricate appear the threads knitting science, technique, economics, and politics together at this time into one pattern of transformation of culture. The period is a crucial one for the development of humanity. It was then and only then that the decisive turn was taken in man's mastery of Nature in the double substitution of multiple mechanisms for the human hand and of steam-power for the weaker forces of man and animal and the inconstant and localized forces of wind and water. The two basic transformations of the sixteenth and seventeenth centuries which made those of the eighteenth possible were the birth of experimental quantitative science and of the capitalist methods of production. At the time when they occurred they still remained largely separate.[4.101] The major practical use of and stimulus to science had been in the field of navigation, an indispensable adjunct to trade but only indirectly connected with production. Very little of immediate practical use came of the great and deliberate effort of the scientists of the seventeenth century, newly banded in their Societies and Academies, to improve manufactures or agriculture (pp. 450 f.). By contrast the later eighteenth century was to see the coming together of the scientific and the capitalist innovations, and their interaction was to set loose forces which were later to transform capitalism and science and with them the lives of all the peoples of the world.

Although there is ample material and even adequate analysis of the political, the economic, the technical, and the scientific transformations of the eighteenth century, these studies have remained largely separate and the combined analysis of them has yet to be written. It would be impossible to embark on it here; the best that can be done is to attempt to put the scientific development in its place against the economic and political background and to trace how far it was affected by, and itself in turn affected, the other aspects of contemporary society.

THE INDUSTRIAL REVOLUTION

The 'Industrial Revolution' is the name that Engels was apparently the first to give it as far back as 1844,[5.45] though it was later sanctified by A. Toynbee.[5.129] Nothing less than the term revolution can be used for the change of productivity in those fields of manufacture in which it first arose. The output of cotton goods rose five-fold between 1766 and 1787.[5.19*] The consequent effects on trade, agriculture, and population were as definite and almost as rapid. Wherever its influence touched a new country, this was marked by a sharp upward break of earlier production trends.

The Industrial Revolution was closely limited in its place of origin; nearly all its major developments occurred in central and northern Britain and mostly in the near neighbourhoods of Birmingham, Manchester, Leeds, Newcastle, and Glasgow. Though the event itself has all the characteristics of an explosive process set off by a particular combination of circumstances that determined the place and time of its occurrence, it remains the final phase of a sustained increase in production over the preceding seventy years or more. Economically this seems to have been determined by the steady growth of a market for manufactured products, mainly textiles, itself largely a consequence of the extended navigations and colonial developments of the seventeenth century.

COAL AND IRON

The combination of economic and political preconditions for a radical change in production specially favoured Britain. It was there rather than in France that manufacture could develop freely to meet the demand, for both feudal and royal restrictions had been swept away by the revolutions of the seventeenth century. The other peculiar advantage of Britain was, paradoxically, the shortage of wood, the basic fuel as well as the basic structural material of all previous civilization. It was this that had forced the development of the use of the inferior, but far cheaper, *coal* for fuel and later of the more expensive but far better material *cast iron* for structures (p. 412). Their production rapidly increased in the later eighteenth century and both engines and mining and metallurgical methods were vastly improved, thanks in part to a new impetus from science marked by such men as Roebuck, Black, Smeaton, and Watt (p. 582). So also were methods of transport, particularly canals.

THE MECHANIZATION OF THE TEXTILE INDUSTRY

The Industrial Revolution itself did not find its origin in developments in heavy industry and transport; it came and could come only from developments within the major industry of the country, and indeed of all countries up to that time: the textile industry. As both the internal and foreign demand for cloth increased, the old merchant and guild-bound industry of southern England could not expand rapidly enough and low wages and freedom from restrictions drew it northwards. There, first in Yorkshire[5.31] and then in Lancashire, it found added advantages in water-power for processes such as fulling, and coal to help with the washing and dyeing. By 1750 the industry came to deal with a new fibre, cotton. Cotton cloth had been imported from India. When this was prohibited at the instance of the clothiers there was a great urge to make it at home. Raw cotton could be grown in the new American plantations. But cotton called for new techniques and was not bound by the old traditions of wool. It was first worked up in the poor district of Lancashire, eminently suitable on account of its damp climate. There, the demand for yarn soon outran the capacity of the old hand-spinning.

There had been isolated attempts at the use of machinery in the textile trade, and even of power-driven machinery, such as that of the stocking frame and of Lombe's silk mill in 1719. They had succeeded but not spread as they had only a limited market to supply. Here, at last, in the cotton industry, was unlimited scope to substitute machinery for hand work. The great inventions, Hargreaves' spinning jenny of 1764, Arkwright's water-frame of 1769, and Crompton's mule of 1779, made the first real breach in the old hand techniques, first by multiplying the action of the hand and then by the use of power in the primary process of spinning. [5.12.508] The relatively enormous output of these machines led to their extension on such a scale as to stretch the capacity of the small streams

153. Workman seated at a stocking frame. From *The Complete Dictionary of Arts and Sciences*, edited by Temple Henry Croker, Thomas Williams and Samuel Clark, London, 1764.

that drove the mills, and in 1785 the last logical step was taken when Watt's steam-engine was adapted to drive them.

INDUSTRIAL CAPITALISM

The textile revolution, which was later to spread to the weaving side with Cartwright's power loom of 1785, and to include wool and linen as well as cotton, was by no means only a technical one. It was made possible only by the social and economic changes of the early eighteenth century and was itself to give rise to the even greater changes of the ninetcenth. To enable the revolution in production to begin, a priming of both *capital* and *labour* was required, for each of these in their modern form had come into existence in this period. Capital was derived in the first place from the great merchant profits of the preceding century, which had begun to skim the resources of the newly-discovered lands in mines and plantations, both worked by slaves, or from the almost undisguised loot of India.[4.5; 5.119] Labour had to be liberated from the land through the enclosures and, as it was no longer cramped by the guild restrictions of the medieval towns, it had to work long hours for low pay in the mills. At first there was not much of it, hence the incentive to labour-saving machinery, particularly such as could be worked by the unskilled, especially women and children.[5.56] Later, with more drastic enclosures and with the importation of poor Irish, there was labour enough and to spare, and the rush of radically new inventions was replaced by an enormous extension of those already in existence, improved but not transformed.

CONCENTRATION OF INDUSTRY

The market for textiles determined the outbreak of the Industrial Revolution in the particularly favourable circumstances which then obtained only in Britain. At one remove the market for textile machinery and textile processing stimulated the iron and chemical industries, while all of these called for an ever-increasing supply of the universal provider, coal, which in turn provoked new departures in mining and transport. By the mid-century, thanks to Darby's invention, cast iron was available in quantity. The shortage was now in wrought iron, and here the need was met for the time being by Cort's method of puddling introduced in 1784. The scientific and technical aspects of these changes will be discussed later (p. 596), but here it is essential to point out again that they ended the age-old dependence on wood as a raw material and brought the iron industry from the forests to the coalfields, where so much other industry was already concentrated (p. 413).

Concentration, indeed, was a prime feature of the Industrial Revolution. Feudal domestic industry, and even urban guild production, was necessarily scattered over many counties. The new mechanical industry hugged the coalfields from the very start. The new industrial towns – Manchester, Birmingham, Newcastle, and Glasgow – accounted between them for nearly all the new products.* These great and growing manufacturing towns, however, exerted their influence far and wide, on one side by their products, the cheapness of which destroyed domestic industry wherever they reached, on the other by their need for hands and for food.[5.1; 5.2; 5.10; 5.56]

THE AGRICULTURAL REVOLUTION

It was this demand that encouraged the new cash-crop agriculture of the landlords and farmers, who were replacing the peasants and their subsistence agriculture over most of England. The agricultural revolution was a mixture of empirical breeding and crop rotation and mechanization with the beginnings of drill ploughs, horse harrows, etc.[5.12.501] It had been prepared by a few enterprising improvers in the early eighteenth century, drawing on Dutch experience, but did not get under way until industry had created a new market for corn and meat and had, as well, provided first the tools and then the power to carry it out. In itself it marks as radical a change in human affairs as the Industrial Revolution. As it advanced, less and less labour was needed on the farm to produce food, which reinforced the tendency to draw the bulk of the population into the cities. Beginning in England, mechanized agriculture was soon to spread to the newly opened lands of America and then, many decades later, to the more populous agricultural parts of Europe.

Interest in agriculture was not limited to temperate zones. The search for tropical products and possible colonies led to further voyages of

154. Mechanized farming in the United States: the 'Champion' harvester by Whitely of Springfield, Ohio. From Edward H. Knight, *The Practical Dictionary of Mechanics*, Cambridge, Massachusetts, c. 1870.

discovery. These were no longer the semi-piratical ventures of the seventeenth century, like those of Dampier, but properly equipped scientific expeditions in which many nations engaged in polite rivalry. Cook (1728–79), Bougainville (1729–1811), and La Pérouse (1741–88) are the most noted examples. Even the ill-fated voyage of the *Bounty* in 1789 was undertaken with the object of introducing bread-fruit trees from the South Seas into the West Indies.[5.136]

THE CREATORS OF THE INDUSTRIAL REVOLUTION

The Industrial Revolution itself did not, in its first stages, depend on any contribution from science; its architects were artisan inventors whose success was made possible by exceptionally favourable economic circumstances. The central developments of textiles did, in fact, occur without the application of any radically new scientific principle. Their real importance was that they marked the emergence of a new social factor in action. The workman with his small accumulated or borrowed capital was here, for the first time, establishing his claim to change and direct the processes of production, in 'the truly revolutionary way', as Marx called it, [4.5.123] as against the mere domination by a merchant of the production of small artisans through the putting-out system.

STEAM-POWER

Nevertheless, in default of the steam-engine and the virtually unlimited power it provided, the Revolution might have gone no further than speeding up textile manufacture in well-watered districts such as Lancashire and the West Riding of Yorkshire, and have achieved little more than had the analogous technical achievements of China many centuries before. It was the use of the steam-engine for power in the textile industry that joined together the two originally separate strands of heavy and light industry and created that modern industrial complex that was to spread from its origin in Britain all over the world. Now the steam-engine, as will be shown later (pp. 576 f.), is pre-eminently a conscious application of scientific thought, and to that extent science played an essential part in the Revolution.

In return the Industrial Revolution itself was to stimulate and support a new outburst of scientific activity. This was even more closely linked with the problems raised by industry than those of the seventeenth century. Not only in England, Scotland, and France, but, as the century wore on, in Russia, Italy, and Germany as well, the movement towards a conscious utilization of science 'for the improvement of arts and manufactures' spread among the newly risen bourgeoisie, and was even smiled on by a section of the aristocracy and the benevolent despots such as Catherine the Great and Joseph II of Austria. But the interest was different from that of the century before; it was more solidly related to achievements in production and it carried a revolutionary flavour.

SCIENCE IN INDUSTRIAL AREAS: THE 'ENCYCLOPEDIE'

It is characteristic that the scientific revival of late eighteenth-century Britain should come no longer, as in the seventeenth century, from Oxford, Cambridge, and London, but from Leeds, Glasgow, Edinburgh, and Manchester, and most of all from the new town of Birmingham,* which became its most celebrated centre (p. 529). In France, where the analogous process was more and more obviously held up by an out-of-date political and social system, the energy of all advanced minds, in despair of any improvement, ultimately turned to getting rid of it, an effort which contributed to the French Revolution. Its monument is the great *Encyclopédie des Sciences, Arts et Métiers* published in twenty-eight volumes from 1751 to 1772 thanks largely to the labours of Diderot (1713–84) and D'Alembert (1717–83), but in which nearly all the *philosophes* took part. This was the bible of the new liberalism, uniting free thought with science, manufactures, and *laisser-faire*.

BENJAMIN FRANKLIN

The most eminent prophet and forerunner of the new movement was Benjamin Franklin, of whom, far more truly than of Canning, it can be said that 'he brought in the New World to redress the balance of the old'. He was born in 1706, the son of a poor tallow-chandler in Boston, Mass. He was apprenticed to a printer and publisher at the age of twelve and ran away to Philadelphia at seventeen to set up on his own. He was sent on a wild-goose chase to England, where he maintained himself as a printer and managed to acquire a thorough knowledge of contemporary science and politics. In 1726 he returned to Philadelphia, laid the foundations of electrical theory, and invented the lightning conductor, the rocking chair, and the iron stove. In 1743 he founded the first American Philosophical Society. He became Postmaster-General of the colonies and equipped the ill-fated expedition of General Braddock against Fort Duquesne (Pittsburgh) in 1755.

Later he returned to England as agent for Pennsylvania and there realized that he had no option but to work for the independence of the colonies, which the aristocratic oligarchy of Britain could neither appreciate nor govern. He was indeed the first to understand the potentialities of the New World and to start planning for its future, as his work on the

155. Frontispiece of Denis Diderot (1713–84), *Encyclopédie ou Dictionnaire Raisonné des Sciences, des Arts et des Métiers, par une Société de Gens de Lettres*, which appeared between 1751 and 1780 at Paris in 35 volumes, 28 volumes coming out between 1751 and 1772 (see page 526). This set of volumes was an immense analysis of trades, crafts, beliefs and comments on every branch of knowledge, and presented a new liberal outlook. It was looked on unfavourably by authority and criticized as subversive, but its publication brought considerable enlightenment.

156. Benjamin Franklin (1706–90): an engraving on steel by James Thomson after the portrait by Joseph Sifrède Duplessis.

Declaration of Independence and the Constitution bear witness. Too old to fight in the War of Independence, he rendered his last, and in some ways greatest, service to his country as ambassador to France, securing the support which proved decisive in that struggle. It was during his stay in Paris and Versailles that he exercised the greatest influence on the direction of politics and science. Franklin was the Bacon of the eighteenth century; but a Bacon with a difference – no longer the wily courtier or learned judge appealing to princes to establish science, but the man of the people born in a freedom that he was determined to preserve and enlarge. He was in the forefront of science in the new age. He joined heartily in the designs of the *philosophes* and added to them the flavour of democracy and practical common sense which they had lacked.

THE DISSENTING ACADEMIES AND THE LUNAR SOCIETY

Franklin's younger contemporaries in Britain carried his ideas into practice. Although, as has been explained, the Industrial Revolution owed little to science, the men who directed its progress were thoroughly imbued with the scientific spirit. The value of science, now less appreciated at court or in the city, was fully grasped by the generation of northern manufacturers and their friends. They saw that the reason why science had not been successful in the past was because its adepts had not been practical men. Further, for the first time outside the navigation schools, it began to be taught systematically. Despite the neglect of the older universities (barred in any case to dissenters, as most of the new men were) it found a place in the dissenting academies, such as those of Warrington and Daventry. Independent foundations, their success was a measure of the need felt for science, and during the eighteenth century they provided, next to the Scottish universities (p. 512), the best scientific education in the world.

It was in this period, far more so than later in the nineteenth century, that the manufacturers, the scientists, and the new professional engineers mixed together in their work and social life. They intermarried, entertained lavishly, talked endlessly, experimented and associated in new projects. This was the age of the 'Lunar Society' of Birmingham[5.105] and the Black Country which used to meet at members' houses on full-moon nights and counted among its members John Wilkinson (1728–1808), the ironmaster who lived and dreamed iron and was buried in an iron coffin; Wedgwood (1744–1817), the potter; Edgeworth, the genial Irishman full of wild and noble-minded projects for social improvement; the serio-comic radical Thomas Day of 'Sandford and Merton'; [5.107] the poetic but practical Dr Erasmus Darwin (1731–1802) of Lichfield;

Joseph Priestley (1733–1804), of whom more later; the melancholy, indefatigable Scotsman James Watt (1736–1819)[5.40] with his younger compatriot, Murdock (1754–1839), the inventor of coal-gas lighting; and finally, the heart and centre of the whole movement, the wealthy, enterprising, jovial, and hospitable Matthew Boulton (1728–1809),[5.41] the Birmingham button-maker who became, as the first manufacturer of steam-engines, almost literally the prime mover of the Industrial Revolution. As he wrote to the Empress Catherine, 'I sell what the whole world wants – power.'

Closely linked with these by personal ties was the more serious group of the Scottish renaissance of the eighteenth century: the philosopher Hume (1711–76), who provided a link with the *philosophes* of France; Adam Smith (1723–90) with his *Wealth of Nations*, the intellectual father of *laisser-faire* capitalism; Dr Black (1728–99), the originator of the pneumatic revolution[5.99]; Dr Hutton (1726–97), the founder of modern geological theory.[5.52] Others, like Dr Roebuck (1718–94), a medical man turned chemical manufacturer and founder of the Carron Works, the first deliberately planned iron-works, and Dr Small (1734–75), the tutor of Thomas Jefferson, belonged equally to England and Scotland.

Such a combination of science and manufacture was only to be found in the Britain of the late eighteenth century. Its existence marks a period of dynamic equilibrium of technics and science, a transition between a period in which science had more to learn from industry than to give to it and one where industry came to be based almost entirely on science. The interests of corresponding circles in other countries were necessarily more economic and political, for they lacked the solid basis that only the new manufacturers could give. Britain appeared to them as a kind of industrial Mecca and indeed some of the best accounts of British industry come from intelligent foreign visitors, such as Gabriel Jars (1732–69), one of the founders of French heavy industry. It is interesting to note that when it was decided in 1782 to start modern iron-working at Le Creusot, the first large works outside Britain, from which not only the French but also the German steel industry is derived, it was necessary to take in W. Wilkinson, the brother of the ironmaster, to cope with the technical side.[5.27]

RATIONAL CHEMISTRY AND THE PNEUMATIC REVOLUTION

The great new scientific contribution of the period of the Industrial Revolution was the foundation of modern, that is to say, rational and quantitative chemistry. This was an event in the history of science of an importance ranking with the great astronomical–mechanical synthesis of a century before. How it occurred will be told in the next chapter; for the moment it is sufficient to say that it marks the result of the rapid development of the chemical industry, largely as an ancillary to the new large-scale mechanical textile industry, and of the consequent interest of scientists in the problems of matter and its transformations.

The actual clue which made possible a simple explanation of the complexities of chemistry was the study of the new gases, itself closely linked with the experiments on air and vacuum of the previous century and with the development of the steam-engine of its own time. Indeed the rise of chemistry may well be called the result of this 'pneumatic revolution'. As a result of the work of pioneer experimenters such as Black in Scotland, Priestley in England, and Scheele in Sweden, the logically trained mind of Lavoisier brought order into the chaos of old and new facts. Twenty years later Dalton provided an explanation in terms of atoms which securely linked chemistry into the Newtonian material-mechanical scheme, though another hundred years had to pass before the nature of the forces between the chemical atoms could be explained (pp. 621–6).*

THE AGE OF REASON: JOSEPH PRIESTLEY

The effects of science were not limited to the industrial field. Beginning with Franklin, the scientists of the later eighteenth century were predominantly, in England as well as in France, radical and liberal in their ideas. The most characteristic figure of this movement which combined the pursuit of science, philanthropy, and radical politics was Joseph Priestley (1733–1804). Son of a Yorkshire cloth dresser, he was educated at the dissenting academy at Daventry with a view to becoming a Congregationalist minister. He drank avidly the new spirit of enlightenment, which did not lead him to infidelity, as it would have done in France, but to a rational Christianity of a more and more Unitarian kind. This did not recommend him to the orthodox, but his learning and interests brought him in contact with the scientific world and particularly with Benjamin Franklin, who inspired him to write a *History of Electricity*,[5.96] which started him on his scientific career. In 1767 he became a minister at Leeds, where he carried out his experiments on carbon dioxide (p. 622). From then on he received the support of manufacturers and some liberal noblemen. He was, in fact, for the most fruitful period of his life, 1773–80, provided with a house and laboratory by Lord Shelburne. It was there that he made his discovery of oxygen which brought him international fame.[5.51]

Nevertheless for him these scientific pursuits were subsidiary to his main purpose of doctrinal controversy in favour of liberal religion. Priestley's religious views were closely linked with his science. Far from wishing, as Descartes had done, to separate matter and spirit, reason and faith, he sought a pure revelation that would unite them. This revelation he sought equally in scripture and in Nature as the work of the divinity. To his mind the activities revealed by electricity showed matter not to be inert and therefore not intrinsically incapable of sensation. In one sense his thought reaches back to the hylozoism of Erigena (p. 293); in the other, forward to the organismal philosophy of Whitehead. He regarded as *Corruptions of Christianity*[5.11.190] such beliefs as those in the Trinity, the Atonement, predestination, and even the existence of the soul. In the eighteenth century such views had a limited appeal. The French were surprised to find a philosopher who believed in God; the English found Priestley's religion difficult to distinguish from atheism. Yet he firmly believed in Christian morals, 'which are

157. The radicalism that Joseph Priestley supported led a mob, incited by the cry of King and Church, to burn down his house and library. From Samuel Smiles, *Lives of Boulton and Watt*, London, 1865.

none other than the well-known duties of life, greater piety towards God, greater benevolence to man'. It was in this spirit that Priestley supported every form of social and cultural improvement, tending, in his words, to 'the greatest happiness of the greatest number.'

He never took an active part in politics, but, at a time when opinion was hardening against the tendencies of the French Revolution, merely to disagree publicly not only with the doctrines of the Church of England, as by Law established, but also with those of respectable dissenters, was considered tantamount to rebellion if not treason. The gentle and benevolent Dr Priestley soon became a radical republican bogy man. The climax came in 1791 when a Birmingham mob, in defence of Church and King and with the connivance of the authorities, burnt down his house near that city, involving the total loss of his library and laboratory. Even when safe from violence he found himself so shunned by his colleagues for his political views that he emigrated to America, where he died in 1804. Events seemed to have made his immediate mission a failure, yet directly or indirectly his influence was to rise again to inspire the liberal and philanthropic movements of the nineteenth century (pp. 545, 1063 f.).

ANTOINE LAURENT LAVOISIER

Priestley's name is indissolubly linked in the history of science with that of Lavoisier, for it was on the basis of the Englishman's pioneer researches that the Frenchman erected the revolutionary theory that was to make chemistry once and for all a rational and quantative science.[5.74] As a personality Lavoisier dominated late eighteenth-century French science. He was a very different man from Priestley. Though, for both, science was only one, if the main, interest in their lives, there was nothing in Lavoisier that corresponded to the vague religious, radical philanthropy of Priestley. Instead the concern of Lavoisier was with efficient public service and the practical use of science to bring the *ancien régime* up-to-date. Lavoisier showed himself from youth as an extremely competent and confident man. In part this was because he was born rich, the last of a family that had risen step by step through care and good management from postilion to postmaster, to merchant, to notary, to attorney, to the *Parlement* of Paris. Lavoisier himself was to take the final step, short of nobility, and buy a place in the *Ferme Générale*, the small and immensely rich corporation that collected taxes for the king. He could not foresee that it was to cost him his head.

His scientific education was of the best and included mathematics, astronomy, botany, anatomy, geology, and, most important of all, chemistry, under Rouelle (1703–70), the genial demonstrator of the *Jardin du Roi*. Here was a young man of ample means, an easy master of all available knowledge, with an ambition to reduce both science and society to some reasonable order. He undertook his first scientific effort when in 1767, at the age of twenty-four, he went on a tour of France to draw up a geological map and make a survey of its mineral resources. Later he was to be occupied with such problems as the system of street lighting, experimental farming, and many other projects of general improvement as characteristic of the eighteenth century in France as in Britain. Most important of all was his appointment in 1775 to the Gunpowder Committee and his establishment in the Arsenal, where he set up what was probably the best laboratory for the time in the world – Priestley's laboratory could be carried on a tray.

Of Lavoisier's scientific work we shall write later (pp. 623 ff.); here we are concerned with him as an influence in the utilization of science, in which he displayed a mastery that was not to be equalled for many years. In everything he did he showed the operation of an exceptionally clear, orderly, and dominating mind. He was not given to philosophy. Though he opened the vast realm of chemistry to the application of physical and mathematical principles, it was the actual illumination he

158. For generating great heat in the laboratory, burning mirrors and lenses were used. This large burning glass was used by Lavoisier and others at the Académie des Sciences. Note the dark glasses worn by the operator (left). After a contemporary woodcut.

brought rather than his methods that remained. His prosecution together with the other Farmers General was not directed against him personally, still less against science. He suffered for the system with which he had been inevitably and conspicuously identified in the movement of a revolution that ironically he had done so much to further.

Priestley and Lavoisier were only two individuals who typified the upsurge of hope and progress so closely linked with the rapid growth of science and industry. Towards the end of the century more and more men, and, for the first time in history, women too, began to think of the possibility of a world ruled by reason and equality and not by prejudice and privilege. This movement spread widely through Europe and the New World, through Italy, Austria, Prussia, Russia, even to Spain. It is noticeable, for instance, how at this time science, long dormant in Italy, showed a revival of the national genius, as the major contributions of Galvani, Volta, and Avogadro show. They were influenced not only by doctrines evolved from their own experience and aspirations, as in the works of J. J. Rousseau (1712–78), but also by everything that was being learned of the eminently reasonable society of the Chinese, of the virtuous society of India, of the noble Redskins, and by the reports of the scientific expeditions on the simple and happy life of the peoples of the coral islands in the South Seas. Society wisely ordered by philosophers and free from the despotism of custom became the ideal, and everything pointed to a return to Nature (pp. 1054 f.).

Science was one of the major inspirations of the era of enlightened despotism. It furnished paradoxically at the same time a new intellectual tool for criticizing and bringing down the old régime and a means for the practical regeneration of mankind through the use of mechanically transformed industry. It was to liberate an enormous scientific and technical outburst which, in its intensity and its consciousness as well as its high level, produced a greater effect on society than anything that the world had seen before.

8.3 The French Revolution and Its Effect on Science

The French scientists of the last days of the monarchy were deeply imbued with the improving spirit of the *philosophes* – the new regime gave them their chance. In the general sweeping away of feudal vestiges and in the exaltation of reason the new science played a leading role. All the revolutionary governments formally recognized its importance, gave much to science and expected as much from it. Some scientists, like Monge (1746–1818) and Lazare Carnot (1753–1823), were ardent republicans and immediately took charge of economic and even military administration. Others, like Bailly (1736–93), Condorcet (1743–94), and the great Lavoisier, though at first they co-operated fully, could not live down their association with the old régime and were victims of the popular reaction to the invasion of France. The majority occupied themselves with the reform of the antiquated machinery of State and of education on scientific lines.

The first task was the reform of weights and measures and the establishment of the metric system, finally achieved in 1799. It needed a revolution to bring it about, as witness the persistence of the old cumbrous systems wherever the influence of France and French logic did not penetrate. The second great task was the creation of modern scientific education, the first real educational change since the Renaissance. The revolutionaries built systematically on a large scale on a foundation that had already been laid in the dissenting academies of England and in the military schools in France, despite the opposition of the old universities. Scotland was an exception; as we have seen, the Scottish universities were from the first in the foreground of scientific advance. Among the products of the dissenting academies were Priestley and Wilkinson the ironmaster – of the French military schools, the mathematicians Monge and Poncelet, and soldiers like Napoleon and, rather surprisingly, Wellington, after he left Eton. For industry and for war science had become indispensable. The foundation of the École Normale Supérieure, of the École de Médecine, and of the greatest of all, the École Polytechnique, gave models for the scientific teaching and research institutions of the future.[5.121] By choosing only the most eminent men to teach in them they created the type of salaried scientist professor that was, throughout the nineteenth century, gradually to replace the gentleman amateur or the patronized client scientist of earlier times.

The first crop of students of the new educational institutions contained such names as Charles (1746–1823), Gay Lussac (1778–1850), Thenard (1777–1853), Malus (1775–1812), and Fresnel (1788–1827), who were all destined to make significant advances in many sciences. These institutions gave opportunity to the gifted of all classes to gain a footing in science. It is to them that France owed her scientific predominance in the world, which lasted till well into the nineteenth century, until Germany and Britain were to follow her example in providing scientific education.

NAPOLEON: PATRON OF SCIENCE

The Napoleonic period, which followed close on the Revolution, markedly speeded up and concentrated the scientific drive. Although the benevolent despots had patronized science, Napoleon took personal charge of its administration. He often attended sessions of the Académie, took a whole scientific expedition with him to Egypt, and was pleased to order the Abbé Haüy (1743–1822) – the founder of crystallography – to write a text-book on physics. He was, after all, the first ruler and the only important one for more than a century with a scientific education. He had therefore some idea, if only a shrewd bourgeois one, of its utility and saw to it that it gave practical support to his régime and to his armies.

The Napoleonic wars had indirectly a considerable importance for science. The Industrial Revolution was only slowly penetrating France by the turn of the century, but France was a far more populous country than Britain, some 28 millions as against 11 millions, and its industrial output, though less concentrated, was actually greater.[5.67.3] It was therefore quite capable of maintaining the unprecedented strain of sending its armies to fight all over Europe. The British blockade, made possible by technical naval superiority, had little damaging effect at the time – its long-term effect was mainly to destroy France's overseas markets. Where it was felt, in cutting off supplies such as soda and sugar, it promoted the French chemical industry and helped to give France chemical predominance for thirty years. Unlike the wars of more modern times, the Napoleonic wars did not extend into the field of science itself, but served rather to promote the meeting of scientists of different countries.[5.38] Napoleon awarded a prize to Davy for his electro-chemical discoveries in 1808, and Davy did not hesitate to go to Paris to take it and protested against the small-minded people who objected merely because the countries were at war.[5.5]

The developments in Britain in the period of the French Revolution were very different. There, instead of vigorous and drastic innovation, there was an almost desperate clinging to the old forms of Church and State, and a rejection of the liberalizing tendencies of the Whigs. Religious dissent turned from rational deism to emotional Methodism. None of these, however, interfered with the march of industry, provided with greater markets as a result of the blockade of France and the additional urge to produce war materials not only for Britain but also for its industrially backward allies.

159. A cartoon by Thomas Rowlandson (1756–1827): a pneumatic experiment at the Royal Institution. The experimenter (centre) is believed to be Thomas Garnett (1766–1802), assisted by Humphry Davy (1778–1829), holding bellows, while Benjamin Thompson, Count Rumford (1753–1814), stands smiling on the right.

THE ROYAL INSTITUTION: COUNT RUMFORD

Only one effort was made at all analogous to the establishment of the new scientific schools of the Continent: the foundation of the Royal Institution in 1799. This was on the initiative of Sir Benjamin Thompson (Count Rumford of the Holy Roman Empire) (1753–1814), an American Tory, but with the same practical bent as Franklin. An opponent of democracy, he saw the need of efficient public service if the old régime was to survive, and demonstrated it by his management of the Kingdom of Bavaria before it was overrun by the French. There he drove the beggars off the streets and put them into workhouses; investigated economical methods of cooking so successfully that they could be fed for three farthings a day; and turned the army budget from a loss to a profit by devising industries for the soldiers. In the course of this he discovered the laws of transmission of heat and demonstrated how it could be generated by work. Returning to England he saw at once that the Industrial Revolution could not be a success unless there was some means of training a new type of mechanic who could base himself on science instead of blind tradition. For that he persuaded the wealthy to put up the money for an institution under royal patronage for:

... diffusing the Knowledge and facilitating the general Introduction of useful mechanical Inventions and Improvements, and for teaching by Courses of Philosophical Lectures and Experiments the applications of Science to the common Purposes of Life.

It did not long preserve its founder's intention. Its first director was the great scientist but even greater snob and showman, Humphry Davy (1778–1829).[5.5; 5.130] He is best known for his invention of the miner's safety lamp in 1815, a piece of direct industrial research which he undertook without fee. Though intended to prevent fire-damp explosions it was used effectively to work previously inaccessible gassy mines, so that output went up while the number of accidents remained about the same.

160. Challenged by the problem of mine explosions due to the ignition of gases, Humphry Davy designed a miners' safety lamp in which the heat of the flame was prevented from reaching the surrounding air and the gases in it. From a tail-piece to a biographical note on Davy in *Gallery of Portraits*, London, 1833.

Davy's paean in favour of the utility of science in his introductory discourse of 1802, given when he was only twenty-three, well expressed the spirit of the new age. In it we find the following expression of the nineteenth-century credo:

> The unequal division of property and of labour, the difference of rank and condition amongst mankind, are the sources of power in civilized life, its moving causes, and even its very soul.[5.5]

With the combination of science, utility, and sound Tory sentiments, it is not surprising that the Royal Institution became a fashionable centre, as popular as the opera with the nobility and gentry.

To make it more exclusive, even the back door, by which the mechanics had been allowed to climb unobserved to the gallery, was bricked up. But it prospered and provided a unique subsidized laboratory in which a large proportion of the basic scientific discoveries of the first half of the nineteenth century were made. Its teaching was confined to public lectures, and though these attracted one of the greatest scientists of all time, the bookbinder's apprentice, Michael Faraday (pp. 608 f.), who was taken on as Davy's assistant and learned his science there, there was no place for hundreds of potential Faradays whom the England of that time could certainly have produced in as great a profusion as France.

THE POST-NAPOLEONIC REACTION

The great movement of the Enlightenment foundered for a time in the reaction that followed the Napoleonic wars – which had, in their earlier stages, done so much to spread the movement throughout Europe – and in the serious slump that followed in the 1820s. It was in such circumstances that the Industrial Revolution showed its ugliest side of unemployment and pauperization, and the ruling classes, with the spectre of another revolution before their eyes, felt obliged to use material and spiritual forces to the utmost to hold down the mob. Men's eyes turned backwards to a somewhat synthetic 'Middle Ages', and a sentimental romanticism took the place of a rational materialism with its irreligious and revolutionary associations.* There was a temporary decline of interest in science except in Germany, where science was linked with awakening nationalism and the windy transcendental *Naturphilosophie* (pp. 645 f.). The industrial need for science was dormant because of the drop in war orders; and there was less need of it than ever in the administration of the Restoration in France and the Holy Alliance. Nevertheless, this decline was only relative to the enormous activity of the last two decades of the eighteenth century. So much had been done then that science was too deeply entrenched in the new industries for this recession to be either as severe or as long as that at the beginning of the eighteenth century. Nor was the spirit of science easily quelled. In Britain, France, and Germany, despite reaction, scientists and admirers of science formed the spearhead of the renewed movement of liberal reform.

8.4 The Character of Science in the Industrial Revolution

The seventy years from 1760 to 1830, and particularly the thirty from 1770 to 1800, are a decisive turning point in world history. They mark the first practical realization of the new powers of machinery in the framework of a new capitalist productive industry. Once these steps had been taken the enormous extension of industry and science of the nineteenth century was inevitable. The new system was so much more efficient and so much cheaper than the old that no serious competition was possible. Nor henceforth could there be any turning back. Sooner or later the whole pattern of life of every human being in the world was to be changed. The critical transition came as a culmination of changes in technology and economics which reached, as has been shown, a breakthrough in Britain, on the technical side, around the year 1760, and in France, on the economic and political side, thirty years later. The changes were not easily effected; it was no accident that the period was one of unprecedented revolutions and wars.[5.9]

In science, also, the eighteenth-century changes were revolutionary – the term pneumatic revolution covers only one aspect of them. Though they appear in conventional histories of science only as an appendage to the Copernican–Galilean–Newtonian rejection of ancient science, this is only a measure of how the historians themselves are still hypnotized by the classical tradition. The seventeenth century had solved the Greeks' problems by new mathematical and experimental methods. The eighteenth-century scientists were to solve by these methods problems that the Greeks had never thought of. But they were to do more: they were to integrate science firmly into the productive mechanism.[5.7] Through power-engineering, chemistry, and electricity, science was to be henceforth indispensable to industry. The first step had been taken in the seventeenth century with the contributions of science to astronomy in the service of navigation. Nevertheless, science still remained largely

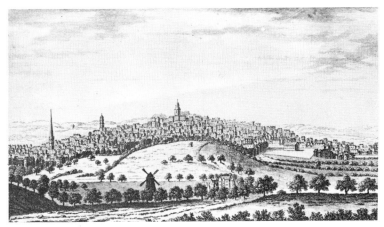

161 a, b. Two contrasted views of Birmingham, one in the second half of the eighteenth century and the second *c*. 1830, showing the expansion of the city and the growth of industrialization.

what it had become in classical times, a somewhat esoteric part of the framework of belief erected in the interest of ruling classes: it was part of the ideological superstructure. Effectively, it had contributed nothing to industry. Now, in the dawn of the nineteenth century, without losing its academic character, it was to become one of the major elements in the productive forces of mankind. This, as we shall see, was to be a permanent feature of growing importance destined to outlast the social forms of capitalism which had assisted at its birth.

In the field of ideas the age of revolutions gave little of importance comparable with that of the scientific discoveries or technical inventions of the period. Time was needed to digest the events and transformations that followed each other in rapid succession from 1760 to 1830. In thought the era lies, as it were, on a watershed. The ideas that inspired the revolutions were those of the French *philosophes* – of Voltaire and Rousseau. They were the heritage of Newton and Locke, based on emotional belief in man and in his perfectibility by free institutions and education once the shackles of Church and King had been loosened. Their German echo was to be found in the profound meditations of Kant (1724–1804), who attempted to weld in one system the achievements of science and the inner light of conscience (pp. 1067 f.).

The ideas that were to come in the nineteenth century were based on the hard experience of the Industrial Revolution and the reluctance of

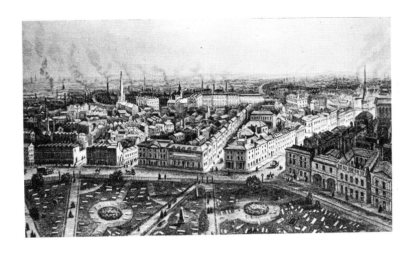

men of culture and property to apply the watchwords of liberty, equality, fraternity too literally. The attempt to apply the social philosophy of the Enlightenment in the French Revolution had revealed serious limitations. It brought out particularly how little the new ideas concerned the lives of the peasants and poor workmen who made up the mass of the population. It was they – the *people* – who had given the Revolution its drive, but when its immediate objects – the abolition of feudal restrictions on private money-making – were achieved, the same people became the *mob*, a threat permanently suspended over the owners of property: the men with a stake in the country. Science, education, liberal theology, from being fashionable, had now become dangerous thoughts. The immediate transition can be seen by comparing Godwin's (1756–1836) optimism with Malthus' (1766–1834) grim and hopeless picture of human existence (p. 1060).

One fundamental advance in ideas was a direct consequence of the great changes of the time. It was the recognition of the historical and irreversible element in human affairs. According to the official, Newtonian, liberal view, Natural Laws, which had been extended from the solar system to cover the world of life and society, were deemed to hold timeless sway. All that was necessary was to discover what those laws were and to arrange industry, agriculture, and society once and for all in accordance with them. The failure of the French Revolution to institute the *age of reason* gave a chance for the alternative view of evolutionary development to gain ground. Vico (1688–1744) had indeed glimpsed this idea in regard to human societies in the early eighteenth century (pp. 1056 f.), and later Buffon (1707–88) and Erasmus Darwin (1731–1802) had speculated that organisms and even the earth itself had had an evolutionary history. It was, however, left to Hegel (1770–1831) to erect these ideas into a philosophical system and for Charles Darwin (1809–1882) and Karl Marx (1818–83) later in the nineteenth century to bring out the consequences of evolutionary struggles in Nature and society (pp. 1067 ff.).

8.5 The Mid Nineteenth Century 1830–70

If in the eighteenth century the curious and far-sighted became aware of the arrival of mechanical industry, by the mid nineteenth century its effects could not fail to be noticed by the most unobservant in every part of the world. Simply by increasing the scale and range of the earlier inventions a complete transformation had been worked in the lives of the tens of millions living in the newly industrialized countries. Vast new cities had shot up filled with rapidly multiplying populations. Beside the growth of industry radically new means of transport had been developed: the railways, which linked up the centres of industry, and the steamships, which collected its raw materials and distributed its products far and wide. Indeed, where the eighteenth century had found the key to *production*, the nineteenth was to find that to *communication*. No comparable change had ever occurred in human conditions with such thoroughness and rapidity. Wherever industrialism spread the old feudal social relations were destroyed. The mass of the population became wage labourers. All economic and political initiative belonged to the new class of capitalist *entrepreneurs*. Even in the State, the relics of feudal reaction had been easily swept away in the success of the revolution of 1830 in France and of the Reform Bill of 1832 in Britain.[5.138] It had become in Marx's phrase 'the executive committee of the ruling class'. It was no longer so necessary to protect privilege by legislation; once property was secure, the workings of the economic system would see to it that everyone got just what he was worth.

Wealth had never been accumulated so easily; misery had never been so widespread and unmitigated by social defences. With all the new triumphs of engineering went a smoky dirtiness, drabness, and ugliness which no previous civilization could have produced. It was in this environment that science approached its present scale of activity and importance. Indeed, as we have seen, already before the beginning of the century it was an indispensable adjunct to the conduct of the new industries, and as the century progressed its range of service continually increased. It grew vastly, and as it did so it necessarily came to be directly influenced by the dominant social forces of capitalism.

It was recognized by the 1830s that a transfer of power from rank to wealth had occurred, even that it might have been necessary. True, in the French Revolution it had exceeded its due bounds, and now that an ideal state of constitutional democracy had been reached there was every reason to resist further fundamental change or even any radical criticism of the abuses of society. In the past science had been a major stimulant to such criticism. Now it was felt by scientists and non-scientists alike that as science was well established its critical and infidel role might well be laid aside.

THE UTILITARIANS

All that was necessary was once more, as in the middle of the seventeenth century, to separate the concepts of science from its social implications: to create an idea of 'pure science' and so, by making science respectable again, to enable it to flourish and, even better, to become really profitable. This transformation was largely effected by the Utilitarians, emasculated followers of the *philosophes* of the eighteenth century. Following the lead of Adam Smith and Jeremy Bentham, they deliberately set themselves the task of removing the old traditional evils of society by legislation which would leave the capitalist absolutely free. It was only thus, under the iron rule of economics as expounded by Ricardo (1772–1823) and J. Stuart Mill (1806–73), that the 'greatest happiness of the greatest number' (p. 533) could be secured. In that age they were superbly confident that the eternal laws of society, as a set of freely contracting independent individuals, had at last been laid bare by science. Firmly believing their new prophets, the entrepreneurs of the golden age of capital set themselves out to prove how right they were. In the enormous burst of productive activity that went on, without any but minor set-backs, from 1830 to 1870 science was to have a small but vital and growing share.[5.54]

This was the period of the heyday of capitalism, with its extravagant wealth and grinding poverty; the period of the Chartists and the Hungry Forties as well as that of the Exhibition of 1851. Capitalism had indeed

162. A Chartist demonstration passing through Blackfriars, in April 1848. The Chartists advocated universal suffrage, the abolition of the property qualification for a seat in Parliament, annual Parliaments, equal representation and a vote by ballot, and the payment of Members of Parliament, all of which constituted, they said, the 'People's Charter'. From the *Illustrated London News*, 1848.

already, as Marx predicted in 1848, brought into existence the dispossessed working class whose eventual power was to bring capitalism's rule to an end. But that day was still far off, and although the fight for better conditions never ceased, increased production and expanding markets did for long enable the capitalists to make timely concessions to the standard of living of the working class.

The mid nineteenth century was not a period of radical technical transformation that can compare with the eighteenth. It was rather one of steadily improving manufacturing methods operating on an ever larger scale. Though rivals were beginning to enter the field, the advantages Britain had won in the Industrial Revolution were retained and even improved on. For a while Britain was literally the workshop of the world. The cheapness of the goods, predominantly the textiles produced by the new machinery, extended the markets in a way that seemed for decades unlimited. That market could be met by simply multiplying and steadily improving existing types of machinery. There was therefore no violent urge for new devices in production. There was, on the other hand, an ever-increasing need to speed up *communication* and *transport*. The *telegraph* was the first practical and large-scale application of the new science of electricity. Materially more important was the application of power to transport in the *railway* and *steamboat*; here science was only in an ancillary role.

THE RISE OF THE ENGINEERS

Both were directly the product of the new profession of mechanical engineers and were made possible by the availability of cheap iron, now smelted with coal, on a scale many times that of any previous age. The appearance of the modern engineer was a new social phenomenon. He is not the lineal descendant of the old military engineer but rather of the millwright and the metal-worker of the days of craftsmanship. Bramah (1748–1814), Maudslay (1771–1831), Muir (1806–88), Whitworth (1803–87) and the great George Stephenson (1781–1848) were all men of this type.[5,113-15] The growth of the applications of science in the mid nineteenth century was so much more rapid than the growth of science itself that their handling and development fell into the hands of practical men. These, for the most part – only the greatest like Richard Trevithick (1771–1833), George Stephenson, and I. K. Brunel (1806–59) were exceptions – proceeded to deal with them as their predecessors had, by trial and error, and to superimpose an evolutionary technical development on the revolutionary innovations that had come directly from science. Thus the reciprocating steam-engine, in spite of nearly 200 years of improvements, is essentially the same machine that left the workshops of Boulton and Watt in 1785.

THE RAILWAY AND THE STEAMBOAT

The railways were originally the products of coal-mining. The great innovation of putting an engine on wheels to make it a *locomotive* also was most successfully attempted in the mines (p. 584). The railway age covered Britain with its network in the thirties and forties and spread to the rest of the world throughout the century. It also led to an enormous increase in the older, civil engineering which carried on the tradition of the eighteenth-century builders of canals, roads, and bridges, like Macadam and Rennie. It can still be seen in the great works of Robert Stephenson and I. K. Brunel. A new interest in geology came from the making of canals and railways, which revealed the structure of the rocks in cuttings and tunnels while at the same time it provided, in the profession of surveyor, a new source of income to the geographical and geological sciences.

THE TELEGRAPH

The improvements in transport brought about by the railway and the steamship put a premium on rapid communications. The need to transmit news rapidly is as old as mankind, as many beacon hills bear witness; but short of magic or telepathy there was little means of realizing it except for alarm calls. Even the needs of war had not produced anything more elaborate than the relay semaphore telegraph. And yet the means had lain to hand for some time. Already in 1737 electricity had been used to transmit messages for distances of several miles, but the use of static electricity was difficult and unreliable. It was the coincidence of the advent of railways with Oersted's discovery of the effects of electric currents on a compass that provided a cheap and foolproof method just when the need was greatest, and ensured the successful invention of the electromagnetic telegraph.

The actual impetus that set a host of inventors working at the same time (e.g. Morse, Wheatstone, etc.) was not any general need of social communication but the actual money value of news of the prices of goods or stocks and of events that might affect them. News meant money, and the electric telegraph provided the means to convey news rapidly.

Short-distance telegraphy was a very direct application of electricity, requiring only a very elementary alphabetic code; but the need for its extension to greater distances and greater speed was to tax the ingenuity of physicists up to the present day and to give rise to much fundamental knowledge and delicate instrumentation. In particular the working of the Atlantic cable, linking Wall Street to the City, was only ensured in 1866 owing to the ingenuity of one of the greatest physicists of his day, William Thomson, Lord Kelvin (1824–1907). Even more important for the general position of science was the fact that the telegraph created the need for trained electricians, and this in turn for technical schools and university departments of physics, on which most of the advances of the later nineteenth century depended (p. 613).

By the fifties science was already paying dividends. A new chemical industry was rising, based mainly on the need of the expanding textile industry for soda and sulphuric acid; and the discovery of aniline dyes secured the future of organic chemistry. A beginning was being made to enlist science, particularly chemistry, in the improvement of agriculture through the use of artificial fertilizers (p. 655).[5.4] Biology was also beginning to find new uses outside the traditional field of agriculture. The chemist, Pasteur (1822–95), was finding means of improving the manufacture of beer and wine and was making his first successful attack on a disease, not of man but, characteristically, of the economically valuable silkworm (p. 649).

Here, for the first time, was the possibility of a scientific, as distinct from a traditional, control of living processes. Even medicine was beginning to move with the times and to accept rather grudgingly such

163. The Great Eastern, designed by Isambard Kingdom Brunel (1806–59), laying the Atlantic Cable in July 1865. On this attempt the cable broke, and it was not until the next year that it was possible to grapple on the ocean bed and recover the cable, splice it, and so forge a telegraphic link between Britain and America. From W. H. Russell, *The Atlantic Telegraph*, London, 1865–6.

gifts as anaesthetics from the new chemistry. Actually, thanks to poverty, overcrowding, and *laisser-faire* economics generally, the health of the people in industrial countries was probably worse than at any other period in their histories. Disastrous epidemics of oriental cholera, brought in through the new facility of transport, continued to occur until their very virulence, and the threat they offered to the middle classes themselves, led to an understanding of the need for sanitation and put some check on the practices of slum landlords (pp. 651 f.).[5.111]

THE ORGANIZATION OF SCIENCE

The facilities for either the practice or the teaching of science by no means corresponded to the function science was already filling in economic life. This was particularly true of England, where science was finding the greatest field of application.[5.8] By 1830 a group of young British scientists under the leadership of Charles Babbage (1792–1871) were especially vocal about the failure of both the Government and the Royal Society, its agent in science, to react to new needs. In his book *Reflections on the Decline of Science in England*,[5.17] Babbage pointed out that the Society had become effectively a closed corporation of officers, controlling a membership the majority of whom had only a nodding acquaintance with science and who were not even its generous patrons. Reform was in the air, but the Royal Society took its time, and by the simple device of restricting its new membership only succeeded in reaching, some years after his death, the state that Babbage demanded.[4.13]

THE BRITISH ASSOCIATION

Babbage was, reasonably, impatient and managed, with his friends, to found in 1831 *The British Association for the Advancement of Science*; a substitute which could be counted on to act and speak on behalf of science. This was modelled on the *Deutsche Naturforscher-Versammlung*, founded in 1822 in Germany by Lorenz Oken (1779–1851), one of the most ardent and fantastic of 'Naturphilosophen' (pp. 645 f.) but a staunch liberal who gave up his chair at Jena in 1819 rather than submit to censorship of his magazine *Isis*. The movement he had started, was, in fact, to be the herald of the great scientific renaissance of Germany in the mid nineteenth century.[5.122]

The British Association was, in its way, as successful. It rapidly became an institution which, though never as august, was far better known than the Royal Society from its habit of carrying its meetings to every city in the United Kingdom and even to the colonies. These meetings were the battleground of all the great scientific controversies of the time, notably those of the conflict of science and religion, culminating in such events as Huxley's retort to Bishop Wilberforce at Oxford in 1860 and Tyndall's Belfast address of 1874 suggesting that life might have come from inanimate matter. It was in part a society for popularizing science, and in part one for promoting and financing research in the interest of the nation. It undertook, for instance, to further the study of seismology, tides, meteorology, magnetism, electrical standards, geology, and biology. Effectively it did by private enterprise what elsewhere was the concern of government. By the end of the century the burden became too great, and was at last shifted by the creation of such institutes as the National Physical Laboratory. One of the Association's actions that was to have the largest consequence was the request it made to Justus von Liebig (1803–73) to prepare a report on agricultural chemistry, a task which turned that great chemist's attention to the practical problems of food production and which was the starting point of the sciences of soil chemistry and nutrition (pp. 633, 655 f.).

Such activities represented the need of the new industrial bourgeoisie to take science into its own hands and to break into the exclusive upper-class and university circles to which it had returned in the early decades

164. The Crystal Palace, venue of the 1851 Exhibition. Built of iron and glass in Hyde Park, London, it was taken down and re-erected at Sydenham, 1852–3, and re-opened in 1854. Designed by Sir Joseph Paxton (1801–65), it was the largest iron-framed building ever constructed, with a nave 1,608 feet long, and a central transept 175 feet high.

of the century. By the mid century they had been largely successful and the new importance of science had obtained institutional recognition.

THE SCIENTIFIC SOCIETIES

The general societies, which had sufficed the seventeenth and eighteenth centuries, could not cope with the flood of specialized knowledge that was creating new fields of science. In France, England, Scotland, Germany, and elsewhere, chemical, geological, astronomical, and other societies were founded, each with its appropriate journal, while at the same time engineers began to associate themselves into institutes.

SCIENCE IN THE UNIVERSITIES

It was also in the mid nineteenth century that the opposition of the English and French universities to the new science, maintained for over 200 years, began to break down. In England this took place partly by the setting-up of new colleges, later to become universities, in London and in manufacturing towns, and partly by adding new departments to already existing universities.[5.128] While at the beginning of the century many, if not most, great scientists in England were amateurs or had been

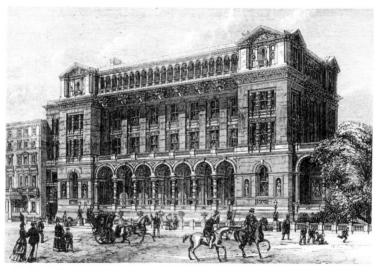

165. The 'New Science School', Exhibition Road, South Kensington. Now part of London University, it was built in 1872 out of the profits of the 1851 Exhibition. From the *Illustrated London News*, 1872.

brought up as apprentices, as were Davy and Faraday, by the middle of the century the university professor, already well known on the Continent, began to be the type of the scientist in England.[5.63] The Great Exhibition of 1851 was a symbol of the unity of science, invention, and manufacture, and some of the proceeds went to founding a scientific teaching centre, the Royal College of Science, in South Kensington. In France the decisive step had been taken much earlier with the establishment of the École Polytechnique and the Ecole Normale Supérieure (p. 536).

It was pre-eminently Germany that took the lead in assimilating science into regular university life. The German universities had indeed begun their reform in the period of the Enlightenment in the eighteenth century. The lead was taken by Göttingen, founded in 1736 by George II in his Hanoverian dominions. From the eighteen-thirties onwards the universities of the diverse German States vied with each other in the foundation of scientific chairs and, though more slowly, of teaching laboratories, of which Liebig's at Giessen was the prototype. Germany had come late into the scientific movement; it had a more disciplined and less independent official class than France or Britain. It was able, however, to supply by organization what it lacked in individual initiative. By the middle of the century, and increasingly thereafter, Germany turned out trained scientists, text-books, and apparatus to supply needs far beyond its boundaries.

All these changes resulted in a vast increase in the volume and prestige of scientific work. It acquired a more and more formal organization, and its pursuit became a profession comparable to the older professions of law and medicine. In doing so, however, it lost much of its early independence, its amateur status. Science did not so much transform the universities as the universities transformed science. The scientist became less of an iconoclast and visionary and more of a pundit, the transmitter of a great tradition. Particularly in Germany, where scientists had first been associated with the liberal movement, they became, after the fiasco of 1848, among the staunchest supporters of the official State machine.[5.3]

MIDDLE-CLASS AND POPULAR SCIENCE

Science was to remain for many years the monopoly of a select part of the middle classes – the liberal intelligentsia as they were known in Europe – and inevitably continued to be limited and coloured by their world outlook. In the middle of the nineteenth century they did not scorn utility. They were interested in the great individual movements of their time. They believed firmly in the inevitability of Progress, but they repudiated all responsibility for any of its unpleasant and dangerous results. Nevertheless, though advancing in wealth and authority, their relative political and economic status had fallen. Industry and finance had advanced in power far more rapidly than science. While in the eighteenth century the leading scientists were on dining and marrying terms with the captains of industry, relatively few could – or seriously wanted to – reach the seats of wealth and power in the nineteenth.

Indeed, for all its growth and extension in the nineteenth century, science only managed to penetrate fitfully, either up or down, beyond the circle of the middle classes. Count Rumford's efforts at the turn of the century to found an institute for the training of mechanics had resulted, after a few years, in a Royal Institution for the scientific entertainment of the nobility and gentry, and only incidentally in the creation of a brilliant research laboratory as well. Other mechanics' institutes succeeded better in their original objective, notably one founded in the City of London in 1823, from which Birkbeck College was to emerge (pp. 1066, 1150). These, however, and the improving lectures given by eminent scientists like Thomas Henry Huxley to the lower orders touched only an insignificant fraction of the new working class which the Industrial Revolution had called into existence. As to technical education, it hardly existed in Britain, the home of mechanical industry, until the twentieth century.[5.3] Those who did not, or could not, resort to 'self help' as a way of getting into the middle classes were apt to regard science and technical innovation generally as a means of cutting wages and producing unemployment (p. 558).

The vision that the new powers of science would make it possible for the working class to get rid of the oppressive system of capitalism, foreshadowed in the pioneer experiments of Robert Owen, was first clearly enunciated by Marx in the *Communist Manifesto* and later elaborated in *Capital*; but the full impact of this doctrine was not to be felt till the next century (pp. 1177 f.).

8.6 The Nineteenth-Century Advances of Science

The mid nineteenth century registered progress in science over such a wide front that it is impossible in a few pages to do more than pick out its major achievements. Physics, chemistry, and biology all expanded and proliferated into separate sub-sciences. There was a great search over all the fields of Nature and technique, such as Bacon had dreamed of but not carried out, by minds already trained in the disciplines of observation, experiment, and calculation bequeathed by the seventeenth and eighteenth centuries. All the previously developed fields continued to deepen their analysis and to find new outlets in practice.

THE TRIUMPH OF CHEMISTRY

Chemistry was especially *the* science of the nineteenth century. This was essentially because it was the major science ancillary to textiles, which all through the century was the most important industry. As will be told in its place (pp. 617 f.), chemistry grew on the secure basis of the revolutionary establishment of the atomic theory and rapidly came to be able to deal with all types of substances. What is important here is that as the century passed, chemistry came to colour in a literal as well as a figurative sense all products of manufacture. New cheap synthetic materials – adulterants, perfumes, dyes, largely drawn from coal tar – came to replace those natural products too dear and rare to cover the new markets. It was in this transition that the centre of chemical research moved from its birthplace in Britain in the eighteenth century, through France, where it was codified and enlarged, to Germany, which was the first country where its multifarious uses were realized in practice. This transition was to prove of sinister consequence in the next century.

THE CONSERVATION OF ENERGY

Amid this active advance of science, old and new, two great generalizations stand out as the major contributions of the nineteenth century. One, in the field of physics, was the doctrine of the *conservation of energy*; the other, in the field of biology, was that of *evolution*. The former, as we shall see (pp. 587 f.), represents the realization, by a whole host of scientists from Carnot to Helmholtz, of the importance, as a cosmic principle, of the interchangeability of different forms of energy. Effectively, it was inspired by the study of the conversion of coal to power that had already been achieved in practice by the steam-engine from the dawn of the Industrial Revolution. It was given a more and more mathematical form and emerged as the science of *thermodynamics*, the first law of which, the conservation of energy, is coupled with a second law which determines its limited availability. It is characteristic of the time that the second law should have been discovered by Sadi Carnot as early as 1824, for it is this law, and not the first, that limits the amount of work that can be got from each ton of coal by an engine of given design. This *efficiency* of engines at that time rarely rose to as much as five per cent.[5.3]

The first law of thermodynamics provided a principle of unification by showing that the forces of Nature previously considered separate – material movement, sound, heat, light, electricity, and magnetism – were all measurable in the same units, those of *energy*, the quantity of which in the universe neither increased nor decreased. Its formulation recalls over the centuries the dictum of Heraclitus (p. 174) on change 'as gold for goods and goods for gold' and is indeed the physical expression of the principle of free trade, established in practice at that very time. The conservation of energy was a magnificent extension of Newton's principle of conservation of motion but, like it, contained in itself no conception of progressive change. Change did indeed follow from the second law, but in the form of degeneration rather than progress, for it showed how in any closed system the heat and the cold must ultimately come together in a uniform tepidity from which no more energy could be extracted (p. 589).

EVOLUTION

Such a conception accorded ill with the progressive and optimistic mode of the nineteenth-century bourgeoisie, who found a congenial scientific justification in the theory of *evolution*. The idea that the earth had a long history was not new. Indeed, as we shall see (pp. 640 f.), it began to take form in the eighteenth century and its acceptance was held back only in the reaction of the early nineteenth century by clerical prejudice. With it came the realization that animals and plants were once very different from what they are now, and the obvious implication that the latter might be descended from the earlier forms. The evidence, however, that was accumulating all through the nineteenth century, derived from the experiences of the age of canal and railway building, made any other explanation very hard to believe. At the same time greater knowledge of the distribution and classification of living animals and plants made the idea of special creation appear more and more arbitrary. Nevertheless it took years of patient and obscure work by generations of geologists and biologists before the world could be made to listen to and begin to accept the idea of organic evolution, with its stinging corollary that man was descended from animals. It needed all the insight, skill, and scientific reputation of Charles Darwin to secure a hearing even as late as 1859 for such a radically new idea with the publication of the *Origin of Species*.

From the moment it was propounded the theory of evolution became the centre of a scientific, ideological, and political battle. Darwin had, almost unwittingly, made as damaging a break in the Platonic doctrine of ideal forms in the animate world as Galileo had in the inanimate. And

166. Charles Darwin (1809–82), from a photograph taken late in life. From Francis Darwin, *Charles Darwin*, London, 1902.

Darwin did more than assert evolution: he provided a mechanism – *natural selection* – that destroyed the last justification for the Aristotelian category of final causes. No wonder the theologians, whose whole world-picture was finalistic, repudiated it. Even more shocking was the idea that man himself – that unique end to creation – was nothing more than a remarkably successful ape. This seemed not only to shatter the doctrine of religion but also the eternal values of rational philosophy. Both were to recover from the blow only too easily (p. 662).

At that time, however, evolution was at the centre of the battle between progress and reaction. For the doctrine found supporters as well as enemies. It was a weapon in the hands of materially minded industrialists against sentimental Tories on the one hand, and idealistic Socialists on the other. It seemed to give a scientific blessing to the exercise of unfettered competition and to justify the wealth of the successful by the doctrine of the *survival of the fittest*. As Darwin's views continued to gain ground and won the enthusiastic support of a new generation of scientists, science itself began to take on again a radical tone, but as yet it was far from a socialistic one (p. 663)*.

The prevailing school of thought following John Stuart Mill, Auguste Comte (1798–1857), and Herbert Spencer (1860–1903) tended to justify, in terms of logic and science, the freedom of private enterprise and to praise the nineteenth century as the era in which man had at last found the right way (pp. 1083 f.). It was not perfect yet; there were still some abuses of the past to be swept away; and progress would continue; but that progress was envisaged as a direct extension of the present – more machinery, more inventions, more accumulation of wealth, even more comforts honestly earned by the deserving poor following the gospel of 'Self Help'. Samuel Smiles (1812–1904) who coined that phrase in his series of biographies of the makers of modern industry showed a sense of significant history far ahead of his contemporaries. Although associated with the doctrine of rugged individualism, he had, towards the end of his life, realized that something more was needed than 'self help', and became a pioneer of technical education for the workers.[5.115]

THE RISE OF SOCIALISM

What the poor thought about the benefits of progress was shown in the Chartist and other revolutionary movements of the mid century and, at the end of the phase, by the insurrectionary Commune of Paris of 1871 following the miseries of war and siege. Their philosopher, Karl Marx, of whom more in his place (pp. 1067 f.), was barred out of the consciousness of the comfortable intellectual classes. Nevertheless, the more honest of these could not help using their eyes and their noses to realize that there was something desperately wrong at the very heart of nineteenth-century prosperity. Artists, poets, and writers were moved to protest against the horrors of the new industrial towns, against the universal degradation of beauty, against the vulgar flaunting of wealth. In opposing them these intellectuals found their first support in an attempt to return to an idealized 'Middle Ages'. Keble (1792–1866) and the Oxford movement, Ruskin (1819–1900) and the Pre-Raphaelites,

167. *King Cophetua and the Beggar Maid* by Sir Edward Burne-Jones (1833–98), a typical example of the medieval idealism of the Pre-Raphaelites.

marked the first reactions which, towards the end of the century, were to become part of the full-blooded Socialism of William Morris (p. 1096).

SCIENCE AND CULTURE

Rejecting industrialism, the literary and artistic movement also largely rejected science, which they felt, with some justification, had identified itself with machine production and all that it had brought in its train.[5.39] It was from this period in mid century that the split between the humanists and scientists, which is such a feature of our own times, first became serious. Its immediate effect was to prevent the co-operation between the two branches of intellectuals without which no constructive criticism of the economic and social system was possible. The humanists never knew enough of how it worked to have other than ineffective emotions about it; the scientists were blunted by a quite deliberate turning away from everything – art, beauty, or social justice – that did not come within the purview of their, by then, highly specialized work.[5.21; 5.117]

8.7 The Late Nineteenth Century 1870–95

Already towards the end of the sixties the first, simple, optimistic phase of early capitalism was beginning to draw to an end. The great depression which started in the seventies marked a transition between the era of free-trade capitalism, with Britain as the workshop of the world, and that of a new, more widely based, finance capitalism, with France, Germany, and the United States coming to the fore under the cover of protected markets. The enormous productive forces liberated by the Industrial Revolution were by then beginning to present their owners with the problem of an ever larger disposable surplus. This could not, under capitalism, be returned to the workers who made it. When invested at home it led to even greater production and to a more hectic search all over the world for markets that were soon filled. The result was colonial expansion, minor wars, and preparations for the larger wars which were to come in the next century.

As a transitional phase it is a difficult period to demarcate, particularly in science. It is certainly easier to do so in retrospect than at the time, for the change here was a gradual one without any marked break in continuity. To those living through that period it seemed that science was moving faster and faster. Nevertheless, doubts had begun to appear as to whether its use was leading to a future of unlimited and beneficial progress. Looking back we see the later nineteenth century as a period which was at the same time an ending and a beginning, a quiet winding up of the great scientific drive of the Newtonian period and a preparation for the stormier scientific and political revolutions of the twentieth century.

In industry, too, the period was transitional. While the old industries continued to expand more slowly in Britain and very rapidly in Germany and the United States, a change was coming over their character. Competition between small family firms led to the formation of big joint-stock companies, soon to become the giant monopolies of the twentieth century. This transition was particularly marked in the metal and engineering industries, in which science was being brought in again after a long run

168. Science and technology were applied to war by the design and manufacture of more destructive weapons; in particular the newly developed techniques of heavy engineering were put to use in this way. Captain Sir Alexander Moncrieff designed this barbette gun carriage in 1868. It allowed the weapon to be fired above the parapet and then loaded in the comparative safety below it. The gun sank down by the force of its recoil, and this 'disappearance' system of mounting heavy ordnance was widely adopted. From the *Illustrated London News*, 1868.

of practical men, and even more in the new chemical and electrical industries which owed their origins entirely to science. With their growth there appear for the first time the Kelvins, the Edisons, the Siemens, the Brunners; not business men turned scientists, but scientists turned business men.[5.3]

We also see, for the first time, a large-scale application of science to war: the submarine, the torpedo, high explosives and big guns, the beginning of the mechanization of warfare. The major industrial characteristics of the end of the nineteenth century were the advent of cheap steel and the introduction of electric power. It also marked the beginning of the use of the internal-combustion engine, which was to revolutionize the transport of the next century. Not less important in their ultimate consequences were the first successes of scientific medicine in reducing the toll of infectious diseases and permitting the exploitation of tropical areas.

THE AGE OF STEEL

The first step in the use of science to transform the traditional iron industry came with Bessemer (1813–98), himself a scientifically minded manufacturer quite outside the industry. His converter, introduced as early as 1854, showed that steel could be made cheaply on a large scale; but the converter was still of only limited use because it required high-grade ore. It was not till 1879, when Gilchrist Thomas introduced the basic-lined open-hearth furnace, that low-grade ores could be used for steel-making and production leaped upwards (pp. 597 f.).[5.124] What is more significant for history was that this changed the geographical centre of gravity of heavy industry. With the basic open-hearth the great phosphatic ore deposits of Lorraine became available for steel-making. In 1870 these had been united with the coal of the Ruhr through the success of the freshly industrialized Prussian State in its war against France.

THE RISE OF GERMAN INDUSTRY

From now on there would be in Europe a centre of steel production which was soon to equal and surpass that of Britain, and on this steel was based a new industry better organized and more closely linked with the State than hers. Britain, however, with her multifarious and competitive industries, still kept a leading, though diminished, place in world markets, particularly by reason of the hold she possessed over all the undeveloped parts of the world.

Rivalries were inevitable and were to be the prime cause of the wars of the next century. In the first stages they expressed themselves, largely owing to the availability of cheap steel, in such exports of capital as rails, locomotives, and agricultural and mining machinery for opening up new territories. These supplemented the still expanding sales of cloth, trinkets, small arms, and hardware, on which the mid nineteenth-century colonialism had been based. What was left of the steel and particularly of the newly developed alloy steels went into battleships and big guns (pp. 705 ff.).

THE ELECTRICAL INDUSTRY

Electricity, as we have seen, played a vital part in the revolution in communications of the mid nineteenth century. The generation of electricity by mechanical force and its use for power transmission were quite evidently feasible (pp. 612 f.) after Faraday's discovery of electromagnetic induction and his demonstration of the electric dyna-motor in 1831. The reasons it was not used for fifty years were, as will be shown, not mainly technical but economic.[5.3] Mid nineteenth-century industry ran on relatively large concentrated units of power – stationary steam-engines for factories, the locomotive or marine engine for traction. The only way of transmitting energy over large distances was by the shipping of coal. Later, the increasing mechanization of minor industries was to call for smaller power units than steam could conveniently supply. The solution

169. The first example of electric traction was the Werner Siemens electric railway at the Berlin Exhibition of 1879. Fed by a conducting rail, the system proved the practicability of this means of transport and led to the development of electric tramways as well as electric railways. From H. Greenly, *Model Electric Locomotives*, London, n.d.

was first found in the gas-engine, the first practical internal-combustion engine and the forerunner of the oil and petrol engines that were to revolutionize transport in the twentieth century.

The electric motor was to prove a far more flexible means of satisfying the industrial need of small static power units. Its whole value, however, depended on the availability of a widespread network of electric power supply, and this could only be brought into existence by a more general need than that of industrial demand. This need was to come from the evolution of domestic services. As the century progressed, extended networks of water and gas supplies, and later of telegraphs and telephones, were laid down. It was an enterprising telegraph clerk, Thomas Alva Edison (1847–1931), who jumped ahead of other competitors and led the way for another such extension – the electric light (p. 613).

Once electricity had to be *made* and *distributed* for light, it could be *used* for power, and a new universal and cheap means of distributing energy was made available to industry and transport, though this was not to be fully effective until the twentieth century. These developments created the heavy electrical industry, which, in contrast to older industries, was monopolistic and scientific from the outset. It was closely linked with other growing monopolies in heavy engineering and with telegraph and telephone monopolies. For science it had another capital importance: it created the industrial research laboratory. Edison's Menlo Park, originally just a barn for trying out inventions, showed the necessity for continuous experiment closely related to production.[5.102]

SCIENTIFIC MEDICINE

While these advances were transforming the manipulable material environment of man, one of even greater importance was taking form: the beginning of a scientific medicine. The reason that this did not take place until such a late date was because the constitution of living organisms was so many times more complicated than that of the most complex mechanical or chemical system that these had first to be understood before a successful attack could even be launched.

Medicine had existed as a mystery and a profession from the very dawn of civilization, but, despite all the progress in the knowledge of anatomy and physiology in ancient and modern times, the doctor could do little more than alleviate the pains and anxiety of the patient and forecast more or less accurately the course of the disease. As human beings recover naturally from most diseases the doctor's care was usually rewarded. The formidable array of drugs in the pharmacopoeia had been compiled partly from the simples of ancient medicine, based on a mixture of folk medicine and magic, and partly from the more violent metallic drugs introduced by Paracelsus in the Renaissance (p. 398). Almost all of them were useless.*

Here and there, for instance in the use of quinine for malaria and of vaccination against smallpox, a few specifics of preventive measures had been hit upon only by fortunate accidents, but for lack of adequate experiment or theory it had been impossible to generalize them. As will be told later, discoveries arising initially from the application of chemistry to the old biological industries of brewing and wine-making led to the first understanding that killing diseases such as anthrax, hydrophobia, cholera, and plague were the result of the invasion of the body by living organisms from outside; and revealed at the same time the way to combat infection by them and, even better, to prevent people from catching them (pp. 651 f.).

From now on, in principle at least, the way to the conquest of disease was open. In its earliest stages it showed that man himself could, through the use of science, overcome what had always before seemed the blind malevolence of fate, or an inscrutable providence beyond his control. In this alone, science had justified itself. Yet the very advances of the new medical science now brought into ever sharper relief the conditions of industrial or colonial poverty that underlay and supported the civilization that seemed so rich and powerful on the surface. The root causes of disease were not in the germs but in the conditions that enabled them to breed and spread, and no vaccine or serum could deal with this evil, which was endemic in the economic system itself.

THE RACE FOR COLONIES

By the end of the century industrialized Europe, largely concentrated on the coalfields that ring the North Sea, had so multiplied in population that it was no longer directly self-supporting. Ever-increasing quantities of food and raw materials had to be imported from eastern Europe, particularly Russia, and from America. It was this demand that led to a rapid transformation in methods of agriculture and of the preservation and transport of food. The development of agricultural machinery, while it did not usually increase the yield per acre, enormously increased that per man. It was especially applicable to open country with a small population, and hence to America, rather than to the older, still feudal village cultures of eastern Europe and Asia.

The introduction of agricultural machinery and the associated rail and steamer transport radically altered the relation of man to his food supply. Before then, even after the eighteenth-century improvements, some eighty to ninety-five per cent of the food produced was consumed on the spot; the town workers and the idle rich, always a small minority, could dispose of only the remaining five to twenty per cent. Countries that lived by trade, like seventeenth-century Holland, or by manufacture, like nineteenth-century Britain, could maintain large urban populations only by drawing on the small individual agricultural surplus of millions of peasants throughout the world. Now, the land workers, using agricultural machinery, could be a dwindling minority and yet provide a hitherto unimaginable surplus for the towns. While this was true at first of grain foods, the principle of urban concentration of food could be extended to meat and fish only by the introduction of refrigeration and canning, involving much physical, chemical, and biological research and development.

These methods of mechanical exploitation, applied mostly to virgin land, had much in common with the mining ventures that were widespread in that period, but, covering a larger area, they were much more destructive in their effects. Exhaustion of the soil was only partially palliated by the use of artificial fertilizers, and the way was open to the devastating erosion of the next century.

170. Transport difficulties led to food wastage in many parts of the world. Here sheep are being boiled down for their tallow alone, since with the great distances in Australia the meat could not be preserved for shipment and was therefore valueless. From the *Illustrated London News*, 1868.

In the opening up of Western and Eastern lands, first to agricultural, then to industrial exploitation, which was largely made possible through the use of steel for agricultural machinery and transport, the financial capital of the older countries found its most profitable outlet. The fate of these investments in these two areas was to be very different. North America, from its foundation a colony of the bourgeoisie, was, even before the Civil War, producing its native capitalists, who were growing rich on the previously untapped resources of the continent and on the labour of tens of millions of poor emigrants from Europe. The Duponts, Astors, Rockefellers, and Morgans were soon to surpass their European forerunners in wealth and power and to turn the United States into the citadel of capitalism.

In Russia, on the other hand, autocracy and the relics of feudalism, combined with an intense exploitation by British, French, and German capitalists, held back development for a time, but when they were swept away in the Revolution the way was open for the first Socialist State.

In the East, India remained for direct, and China for indirect exploitation; but one State – Japan – was allowed to become a model of the civilizing value of a native capitalism, creating all the appearances of the new 'Western' culture, including science, but using them to build up on a feudal basis an uninhibited, predatory and militarized State.* [5.131]

8.8 Science in the Late Nineteenth Century

In a period so short and so crowded with practical achievement as the late nineteenth century it is not to be expected that many great theoretical advances would be made. In the physical sciences the period was pre-eminently one of transition, with the rounding off of the great advances of the early nineteenth century, and at the same time with the beginnings of investigations of a new kind that were to lead to the explosive advance of the twentieth century. In biology, on the other hand, new ground was being broken in the study of microbes and in the approach to a physical–chemical understanding of physiology.

THE ELECTROMAGNETIC THEORY OF LIGHT

The major achievement of the period in physics was the formulation by Maxwell of the *electromagnetic theory of light*. This brought together in one comprehensive theory the results of two generations of experiments and theories in different fields of physics – electricity, magnetism, and optics – and gave them a simple mathematical formulation. Though in itself a triumph of mathematical physics, it depended for its verification on the establishment of accurate units for electricity, a task made necessary by the rise of the electrical industry. In turn, Maxwell's equations were to form the theoretical basis of future electrical engineering, an intricate interplay of theory and practice.

The electromagnetic theory was a crowning achievement which realized the dream of Faraday that all the forces of Nature should be shown to be related, and, together with the laws of thermodynamics, seemed to imply a certain finality in physics – an idea that was to be rudely shattered in the twentieth century. It was, however, also to be a beginning, for its central concept – the theoretical necessity for the existence of electromagnetic waves – was to lead to their experimental demonstration by Hertz in 1888, and from this to their practical utilization in wireless telegraphy and all that was to result from it.

THE PERIODIC TABLE OF ELEMENTS

In chemistry the period included one major generalization, the *periodic table* of Mendeleev (p. 740), put forward in 1869, which also appeared at the time to set a limit to the existence of fundamentally different kinds of matter, but which was actually to find its full interpretation in a new concept of matter no longer made of immutable atoms but by relatively impermanent associations of a few fundamental particles themselves liable to change and transformation. Mendeleev was the Copernicus of the atomic system; its Galileo and Newton were still to come.

In organic chemistry, once the confusions due to a reluctance to accept the atomic theory had been overcome, there was a magnificent and ordered advance in the interpretation of the structures of natural substances and an even more impressive deliberate synthesis of new ones. By the end of the century chemical research was fully embedded as an essential part of the new chemical industry, which was now reaching out from its triumphs in synthetic dyes to those of synthetic drugs (p. 634). The chemists had so multiplied in numbers as to represent well over half of all workers in science.

RESEARCH LABORATORIES

This greater utilization of science and scientists called for large extensions in scientific education and in the organization of science. The only organizational innovation was the advent of the industrial research laboratory, which grew up almost imperceptibly from the workshop

171. The chemistry laboratory for the School of Practical Chemistry, University College, London, in 1846: one of the new and growing university science laboratories. From the *Illustrated London News*, 1846.

or private testing place of the inventor turned business man, such as Siemens or Edison. But university laboratories also grew, from the very fact that the new uses of science meant new jobs and attracted more and more students. Thus, despite all the protestations of disinterestedness, the academic science of the period was ultimately dependent on the success of science in industry. Nevertheless it was left for the most part to the enjoyment of considerable liberty as long as it respected conventional limits in politics and religion.

THE DOMINANCE OF GERMAN SCIENCE

The largest increase was to be found in Germany, which, from the number of its universities, its newly founded *Technische Hochschulen*, and its innumerable *Zeitschriften* and *Handbücher*, tended increasingly to dominate the scientific world towards the end of the century. Britain and France, relying on their own great traditions, resisted this tendency, but the German language became pre-eminently the international language of science, and German professors set up a kind of scientific empire which covered all northern, central, and eastern Europe and exerted considerable influence on the science of Russia, the United States, and Japan. The German professor was on the way to becoming the model for the scientists of all the world. Like most of the German intellectuals, he had made his peace with the alliance of military feudalism and big business which ruled that newly industrialized and expanding State. That allegiance was to point to the next stage in the development of science, in which it was to come to be used in the service of the State predominantly for military ends.

THE GREAT DEPRESSION

The end of the nineteenth century, like its beginning, was marked by a philosophical reaction tending to limit severely the field and significance of science. But whereas the early reaction was directed in opposition to the effects of the French Revolution, the later one was dictated by an anxious awareness of a social revolution to come. In spite of the enormous new wealth that was being produced by industries ever more scientific in their operation, in spite of the prospects of further advance, the strains of society seemed to increase rather than diminish, and there could be no denying a sense of frustration and doom in the ranks of the cultured intellectuals, a *fin de siècle* feeling that was to be only too well justified. Especially in Europe, Marxist Socialism seemed to be providing a hopeful alternative for the industrial working class. It was there, accordingly, that the current of philosophy was most directly affected, but Britain and America, for all their traditional indifference to philosophy, were not immune.

There was a turning back from the implicit and optimistic materialism of the mid century towards the neo-positivism of Mach (1838–1916) and Ostwald (1853–1932), who, under the guise of purging science of unnecessary mental constructs, removed matter and replaced it by bundles of sensations or convenient fictions. This philosophy and others like it, such as the *élan vital* of Bergson (1859–1941) and the pragmatism of William James (1842–1910), all tended to take the revolutionary sting out of science, to laugh out of court any idea that it could be used to effect any significant improvement in the lot of man, and to make it acceptable to organized religion and the State (pp. 1090 f.).

These philosophies were indeed only symptoms of the absorption of science, as a consequence of its growing technical indispensability, into the machinery of capitalism. The change in the scientists' attitude towards pure science and away from social responsibility was made easier by increased endowments, permitting greater specialization, and by a discreet distribution of honours and patronage. The very increase in the number of scientists also reinforced this tendency to conform and escape responsibility. By the end of the century, the independent scientists were a small minority. The majority drew their salaries from universities or from the Government, and more than ever assimilated the mentality of the ruling class.

How far these conformist tendencies held back the development of science will always be hard to tell, because in actual history their influence was outweighed by the enormous expansion in the scale of science itself. But that there was some such retarding effect seems to emerge from all detailed studies of the advance of particular sciences.[5.3] It is not so much that phenomena have been missed or that, when observed, conclusions that afterwards seem to be obvious have not been drawn from them, though this has certainly occurred over and over again; it is rather that in the social system of the late nineteenth century there existed no real sense of direction or idea of the relative significance of different fields of work. Had there been such, many of the great discoveries which were to come at the turn of the century could have been anticipated by twenty years or more. The effort wasted on fruitless refining of old theories would have amply sufficed to have brought the new to light. It may be said that such an idea was alien to science at the time – some say it is so still – but there can be no doubt that the all-round, organized scientific drive of great periods, such as those of the mid seventeenth and late eighteenth centuries, and even of the mid nineteenth century, seemed to have vanished. It was not until the disturbed period of the twentieth century that it was to appear again in full vigour.

This closes the account of the general development of science in the eighteenth and nineteenth centuries. A general appreciation of the achievements of that great era is deferred to the end of Chapter 9, after considering in more detail the progress of the separate sciences.

Developments of the Sciences
in the Eighteenth and Nineteenth Centuries

9.0 Introduction

By the time the eighteenth century is reached the relations of science with society can no longer be set out as one simple time sequence. It was necessary to begin with such a sequence because without it the histories of the sciences remain mere chronicles; but by itself it conceals the inner connexions of particular sciences that run on continuously over the whole period. In each science the parallel growth of understanding and control depends both on inner and outer factors. The inner determinants are the hard facts of Nature: the structure of matter and the events and characters of its evolution. The outer determinants are the technical, social, and economic capacities and drives that link with general history. These, though they may not determine what is found out, are decisive in determining when and how new facts are brought into the cumulative tradition of science. To understand fully how this happens it would be necessary to follow the history of science in detail, with wider knowledge and critical ability than have usually been brought to bear on it hitherto. I can make no claim to do so here, but will try merely to exemplify some of the principles of the interaction by a treatment in outline of some selected fields of science and technology that between them bring out the general character of the advances of the eighteenth and nineteenth centuries.

The fields I have chosen are those of Heat and Energy (9.1); Engineering and Metallurgy (9.2); Electricity and Magnetism (9.3); Chemistry (9.4); and Biology (9.5). In the last part of this chapter (9.6) I have attempted to pull together the material in this and the preceding chapter and to examine the lessons to be drawn from the time and subject sequences. The choice of subjects has been made to bring out the major features of the eighteenth–nineteenth-century transition from a science which was largely academic to one beginning to play an essential part in economic life. In all but the second case, each subject includes one or more economically important development linked with the discovery of some principle of fundamental scientific importance. Thus the first section contains the history of the steam-engine, and shows how attempts to increase its efficiency led to the discovery of the laws of *conservation* and *transformation* of *energy*. The second section is, in a sense, an appendage of the first, for it was the exacting demands of the construction of steam-engines and steam-driven machinery that led to both the precision methods of metal-working and the production of improved metal in quantity, leading to the *age of steel*. Here no great scientific principles are involved and the amount of science called on is relatively small. The value of the study of engineering is that it brings out first how much of the mechanical transformation depended on simple workmen and secondly how essential precision metal-working was, both to industry and science. In the story of steel the emphasis is rather on the enormous technical and economic advances that were achieved by the employment of relatively little scientific knowledge.

In the third section, on electricity, we have a different case again – the study of the transformation of a subject of purely scientific and even frivolous interest into a major industry. At the same time it should serve to bring out how the applications of the mathematical mechanics evolved in the seventeenth century to a field of entirely unexpected experience, were in the nineteenth able to create new generalizations of the greatest theoretical importance. The steps leading to the *electromagnetic theory of light* are comparable with those leading to Newton's theory of gravitation. It represents in itself the second major unifying hypothesis which gave nineteenth-century science its deceptively final character.

The fourth section recounts the central advance of eighteenth-century science which resulted in bringing the field of chemistry, previously shared between blind empiricism and mystical alchemical theory, into the range of rational quantitative science. The *pneumatic revolution* associated with Priestley and Lavoisier represents the first large-scale extension of science beyond the region cultivated by the Greeks. Its overwhelming importance in human history lies in the fact that it was also the first in which science entered, in a positive and profitable way, into a major productive industry. The subsequent close association of chemistry with the textile industries, with the passage from bleach and dyes to explosives and drugs, is the theme that accompanied and inspired *organic chemistry* in the nineteenth century.

Finally, from the vast field of biological sciences I have tried to bring out two or three leading threads which determined directions of advance. Here we have, on the one hand, agricultural and medical preoccupations which led ultimately to *microbiology* and to Pasteur's germ theory of disease. On the other hand, there is the passionate controversy on creation that was to lead through geology and natural history to Darwin's establishment of organic *evolution*. There can be no doubt that of all the great achievements of nineteenth-century science, including the magnificent generalizations of physics, only evolution is comparable in importance to the Copernican–Galilean dethronement of the earth as the centre of the universe. Henceforth man himself found his place in Nature. Only by recognizing that he was an animal could he learn how different from his ancestors the operations of society and civilization had made him. With the acceptance of evolution the last link with the Aristotelian world-picture was snapped; but the logical implications of a man-made world in the place of a celestial providential mechanism had yet to be drawn – a task that was to prove too difficult in the framework of capitalist society.

By focusing attention on the major scientific and technical advances of the period, I inevitably over-simplify the picture and am forced to leave out whole sequences of topics that would be needed for a comprehensive treatment. There is, however, no reason to believe that the story they would tell would be of a different kind. I have said, for instance, little or nothing about the great development of optics that set in at the beginning of the nineteenth century, involving the discoveries of polarization and diffraction that led to the reappearance of the wave theory of light, nor about spectroscopy and spectral analysis. These developments were to multiply the number of instruments available to other sciences, to transform chemistry and astronomy, and, in the next century, to provide a clue to the structure of the atom (pp. 739 f.). The story of optics abounds in examples of the interplay of scientific and economic factors even in the nineteenth century, before the days of the cinema and television, but there is no space to tell of them here. The discussions in the sections which follow should, however, show sufficient of the types of interaction to cover those of the fields which are not treated.

9.1 Heat and Energy

The study of heat and its transformations was one of great intellectual, and even greater technical and economic, importance for the development of modern civilization. Originally, it was merely an extension of observations of Nature, of feelings of warmth and cold, of the operations of cooking, of the changes of the weather. There had been plenty of early speculations about heat. It was clearly connected with both life and fire, as also with violent action.

The Ionian philosophers had, following even earlier legends, brought in heat and its opposite cold as the causes of the evolution of the universe – heat expanding and vaporizing, cold congealing and hardening. Aristotle, especially in his meteorology, fixed the doctrine of the qualities of hot and cold, which, with wet and dry, determined the canonical four elements of fire (hot, dry), water (cold, wet), air (hot, wet), and earth (cold, dry) (p. 199).

This doctrine, a fusion of chemistry and physics, was engraved for millennia in human thought, as much in China and India as in Europe. The doctrine of antagonistic elements was particularly important in medicine and seemed to be supported by the experience of chills and fevers. Indeed it is from medicine that came the first elementary ideas of heat measurement. Heat and cold were supposed each to be ranked in four *degrees* or steps, the first just perceptible, the fourth mortal. The object of heating or cooling medicines of the first, second, or third degree was to correct and temper its opposite, hence the idea of *temperature*.

This philosophical medical doctrine survived and took new life in the Renaissance. Bacon, following Telesius, had even made the antithesis of heat and cold a central feature of his philosophy (p. 442). From the very earliest times heat was associated with the movement of airs and vapours, and it was largely through its connexion with the pneumatic discoveries of the seventeenth century that it left the orbit of qualitative philosophy to enter that of quantitative science. Galileo had constructed an air-expansion thermometer, and such thermometers, together with Torricelli's barometer, were used for observation of the weather.[4.22]

THE EVOLUTION OF THE STEAM-ENGINE

The line of advance in the quantitative study of heat was not, however, to lie through such investigations, but along the practical road of the utilization of the power of expansion to make heat do useful work. All through the seventeenth century the idea of 'raising water by fire' fascinated ingenious projectors (p. 471). The problem was how to combine two ideas, both old, into a practical engine: first to fill an empty space with water by suction (or vacuum), and then to expel the contents by pressure exerted by expanding air, steam, or gas. De Caus (1576–1626), a designer of the garden waterworks so favoured in the sixteenth century, solved the problem practically even before the vacuum was realized. He lit a fire under an almost empty vessel of water connected by a pipe to a well; when it had boiled away and the vessel was full of steam he took the fire away and closed the steam vent, and water was sucked up to fill the empty space. Though scarcely practical, this contained the essential principle of a vacuum engine, but until van Guericke's work (p. 470) its action could not be fully understood. Most of the scientists working on vacua had some idea of a practical engine but lacked the mechanical ability to make one that would work. The man who came nearest to it was Denis Papin, assistant to Huygens and Boyle in succession, who drew up the specifications of such a machine but could not raise the money to make it. He died in poverty in London. We have a pathetic letter from him to the Secretary of the Royal Society in 1708 asking for a sum of fifteen pounds for a 'considerable experiment', with the Society's answer that it could not loan money unless assured of success in advance.[4.19.38]

The man who first succeeded in designing and financing a workable fire-driven pump was Captain Savery (1650–1715) of the Royal Engineers, who used two vessels alternately filled with steam to drive water out and then cooled to draw up more water, a method still in use in the 'pulsometer' pump. Savery was no ordinary projector. He was fully aware, as his Patent Application entitled *The Miner's Friend* shows,[5.104] of the possible importance of the steam-engine, especially for draining mines, where there was the greatest need for heavy continuous work. There he says:

To the Gentlemen Adventurers in the Mines of England.

I am very sensible a great many among you do as yet look on my invention of raising water by the impellent force of fire a useless sort of a project that never can answer my designs or pretensions; and that it is altogether impossible that such an engine as this can be wrought underground and succeed in the raising of water, and dreining your mines, so as to deserve any incouragement from you. I am not very fond of lying under the scandal of a bare projector, and therefore present you here with a draught of my machine, and lay before you the uses of it, and leave it to your consideration whether it be worth your while to make use of it or no . . .

For draining of mines and coal-pits, the use of the engine will sufficiently recommend itself in raising water so easie and cheap, and I do not doubt but that in a few years it will be a means of making our mining trade, which is no small part of the wealth of this kingdome, double if not treble to what it now is. And if such vast quantities of lead, tin, and coals are now yearly exported,

172. The steam-engine of Thomas Savery (1650–1715), designed expressly for pumping out water from mines.

under the difficulties of such an immense charge and pains as the miners, etc., are now at to discharge their water, how much more may be hereafter exported when the charge will be very much lessen'd by the use of this engine every way fitted for the use of mines?

Nevertheless, Savery's engine laboured under several practical disadvantages, and its main value was to show that the problem could be solved. A more successful and practical engine was made in 1712 by the ironmonger Thomas Newcomen of Dartmouth, who used a piston which was depressed by condensing steam in a cylinder connected directly to a low-pressure boiler. Newcomen's engine did not, like Savery's, need to be built at the bottom of the mineshaft, it required less attention, and, not depending on high steam pressure, it was much safer. Its introduction marked the first stage in the translation of the scientific

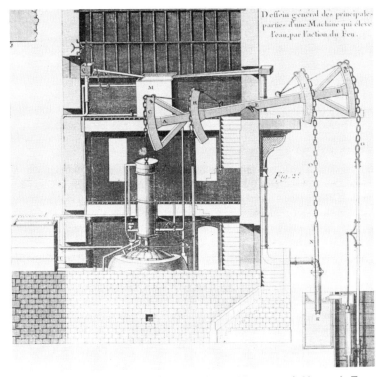

173. The beam steam-engine designed by Thomas Newcomen (1663–1729). From Bernard Belidor, *Architecture Hydraulique*, Paris, 1737–53.

principle of atmospheric pressure into a machine that could be built
by practical men and would not only work but also pay.

The fact that, as far as we know, Newcomen had no scientific training
or connexions [5.12.611] is among the reasons that caused R. S. Meikleham
in 1824 to repudiate the view that the steam-engine was 'one of the noblest
gifts of science to mankind'. 'There is no machine or mechanism,' he
asserted, 'in which the little the theorists have done is more useless. It
arose, was improved and perfected by working mechanics – and by
them only.' [5.84] These two extreme views as to the share of science in the
origin of the steam-engine are not incompatible. It is doubtful whether
the radical idea of vacuum pumping would ever have occurred to a
mechanic, at least it did not before it occurred to a scientist; on the other
hand, no scientist either had or could command the skill to solve the,
no less essential, problems of making a working engine. As the sequel
shows, the repeated combination of radical scientific ideas and experi-
enced craftsmanship was needed in the further development of engines.

It says much for Newcomen's ingenuity that no radical improvement
in his engine was made for nearly seventy years, and that some of the
engines themselves ran for more than a hundred. But it was limited in
its use, its action was too irregular for anything but pumping and blowing, and it consumed enormous quantities of coal. Its further development had to call on an injection of new ideas from the side of science,
in particular from the creation of a quantitative science of heat.

SPECIFIC HEAT AND LATENT HEAT: JOSEPH BLACK

Heat began to become a quantitative science with the gradual expansion
and increase in scale of the industrial operations which made the largest
use of it. It grew out of the scientific appreciation of the experience of the
distillers and salt-makers, who were accustomed to boiling and condensing liquids on a large scale, and then out of that of the makers and users
of the early steam-engines.

Dr Black, whose contribution to chemistry was to set off the pneumatic
revolution (p. 621), was also the originator of the new view of heat. His
approach was in the first place a medical–physical one. He was concerned with elucidating the nature of the element fire or heat that could
pass through vessels and affect their contents. He found that different
substances were heated to different degrees by the same amount of what
he called the 'matter of heat'. This he discovered by the method of
mixtures, which had first been used by Jean Morin (1583–1656)[5.87] – still
working on the Arabic idea of four degrees of heat balancing four
degrees of cold – and carried it to the point of establishing the heat

capacity or *specific heat* of different substances. It was at this point that he reflected on the fact that snow and ice took time to melt – that is absorbed heat without getting hotter – and that the heat must be hidden or *latent* in melted water. He next measured the large latent heat of steam, which is reflected in the fact, long known in the distillery trade, that it requires a very much greater amount of heat to boil water away

174. Watt's double acting steam-engine. Watt patented improvements to his steam-engine in 1782. It was then well-known because it used a separate condenser to cool the steam after use in the main cylinder, thus enabling the main cylinder to remain hot and thereby increasing the efficiency of the machine and reducing fuel costs. The main cylinder is shown on the left of the photograph. His 'double acting' improvement increased efficiency; by means of a mechanical leverage system (shown above the main cylinder) a thrust was given to the beam (shown across the top of the engine) each time the piston in the cylinder moved down as well as up. Steam was also admitted to the cylinder only during the early part of each stroke: this improvement meant that the steam could be used expansively. The conical governor with two large balls (centre) ensured a constant speed with varying loads, since the governor operated a valve to control the amount of steam admitted to the main cylinder. The separate condenser lies within the casing in the lower half of the engine.

than to raise water to the boil. Further, the heat absorbed in boiling was recovered when the steam condensed again in the worm of the still, where the application of much cold water was needed to get rid of it.

JAMES WATT: THE SEPARATE CONDENSER

The first practical application of the discovery of latent heat was to be made by a young Glasgow instrument maker, James Watt, [5.40] who was charged with repairing a model Newcomen engine for the university (note here again the reciprocal action of technique on science). He found that the trouble was due to the steam lost at each stroke by condensing in the cold cylinder. Black gave him the explanation in terms of his newly discovered latent heat, and not long after Watt hit upon the idea of condensing the steam separately. This invention of the separate condenser in 1765 was crucial for the development of the steam-engine, as it made it immensely more efficient. The condenser was only the starting point of Watt's improvements.[5.58]

MATTHEW BOULTON: THE SOHO ENGINE WORKS

Before a saleable engine could be made Watt had, after a relative failure in Roebuck's Carron works, to go into partnership with the great Matthew Boulton, the Birmingham manufacturer (p. 530), and make use of the resources of the growing metal industry of the Black Country before the steam-engine could be turned from an idea into a reality, for, as Watt himself admitted with unconscious irony, 'the Scots were naturally incapable of becoming engineers'. Particularly valuable were the services of John Wilkinson's cannon-boring machinery for providing true cylinders. By introducing the combination of flywheel, throttle, and centrifugal *governor*, Watt made an engine capable of driving machinery at steady speed even against very variable loads. This device in itself is the first example of feed-back or *cybernetic* control in industry (pp. 783 f.). Appearing at the very outset of the great Industrial Revolution, it was a portent of the *automatism* characteristic of the second industrial revolution of the twentieth century.*

Until Watt's time steam-engines were only exceptionally used for mines well away from the coalfields; the Newcomen engine, even when improved by Smeaton (1724–92), was a paying proposition only in pumping coal-mines, where coal was naturally extremely cheap. But with the more efficient and steady Watt engine the products of the whole field of heavy-metal mining in Cornwall and, later, the power for the industrial textile mills, which were spreading all over the country, became easily available and cheap.

175. A general view of the Soho Works, Birmingham, of Matthew Boulton and James Watt where the development of the steam-engine was advanced. From a woodcut published in about 1830.

After a great struggle, for there were many economic and technical difficulties to overcome, the steam-engine won its way into every mining and manufacturing district of Britain. Nor did it stop there, for Boulton believed in manufacturing for the whole world: steam-engines were set up in France, in Russia, and in Germany, most often by engineers from Britain.

THE LOCOMOTIVE AND THE MARINE ENGINE

The subsequent development of the steam-engine was conditioned by the technical and economic requirements it was called on to meet. Watt's engine was satisfactory enough for most mine and factory purposes, but it was expensive, heavy for the power developed, and still used too much coal. Where lightness and high power were needed was for a *locomotive* engine. Here the answer lay, as Trevithick had shown as far back as 1801, in a *high-pressure engine*, dispensing with the condenser altogether and blowing the exhaust steam into the air.[5.42]

The locomotive got off to a fumbling start. It grew up in its natural home in the coalfields on the *railway*, between pit and loading staithe (p. 413). Before it could possibly pay there were innumerable problems to be solved, of drive, suspension, rails, permanent way, so that it is not surprising that here science had little part and that it was the self-taught colliery fireman's son, George Stephenson, who went farthest in solving all of them.[5.99] His decisive invention, made almost by chance, was turning the exhaust steam into the funnel and thus, by urging the fire, getting enough power to beat the horses and reach the phenomenal speed of twenty miles an hour. The acknowledged triumph of the locomotive came with the Rainhills trials in 1829 of the new Liverpool–Manchester railway, where his Rocket took the prize.

The problem of adaptation to water transport was a very different one; there weight and size were not important but fuel economy was, for the steamer must carry its own coal. Indeed this limitation was to confine steamboats to river and coastal trade for most of the nineteenth century. The solution was found in the use of multiple expansion introduced by Hornblower (1743–1815) in 1781, but only slowly developed.

176. The Rainhill locomotive trials: in 1829 the directors of the Liverpool and Manchester Railway offered a prize for the best locomotive, and in October of that year trials were held at Rainhill, four miles from St Helens in Lancashire. Four locomotives took part and the competition was won by George Stephenson's 'Rocket'.

No radical change, except the substitution of the screw for the paddle, was made until 1884, when Parsons' turbine revolutionized power production.

INTERACTION OF ECONOMICS AND TECHNIQUE IN THE INDUSTRIAL REVOLUTION

The history of the steam-engine shows how the necessary conditions for the Industrial Revolution were both economic and technical; economic, in the sense that the textile industry grew up to provide the consumer goods for sale in an expanding market; technical, in the sense that the new engines were the only means of providing the coal and motive power, and ultimately the transport, without which the expansion of the textile industry would have been impossible.

For the most part the steam-engine was improved by practical engineers without any notable contribution from science. Its working did, however, attract the attention of many scientists who wished to understand it or even hoped to improve it. This study resulted in a much deeper appreciation of the laws governing the behaviour of gases and vapours – needed for drawing up steam tables – and was to lead to the new general conception in physics equating mechanical force and heat in theory, as the steam-engine already had in practice, in the common term of *energy*.

THE ESTABLISHMENT OF CALORIC

Paradoxically, it was in France, where the steam-engine was a foreign importation, rather than in Britain, where it originated, that its operation as a means of transforming heat into work was first given serious scientific study. The primary difficulty lay in the traditional ideas that existed as to what heat meant. As we have seen (pp. 575 f.), heat was confused with fire; even the vitally important animal heat was attributed to an invisible fire.[4.132] In the eighteenth century it was thought of as a material substance, Black's 'matter of heat', later to be christened *caloric* by Lavoisier. Though attempts to weigh it failed, this only showed it was an imponderable fluid like electricity or light.[5.75] Lavoisier (p. 625) showed that this concept fitted very well with his idea of the generation of heat by chemical combination, particularly of combination with oxygen in a fire or in an animal body.

Nevertheless, an entirely different tradition, that heat was a form of motion and not a substance at all, also existed and was of even greater antiquity. Centuries of experience with the fire-drill and the forge had shown that force could be turned into heat; the steam-engine now demonstrated that heat could be turned into force. But it also needed the steam-engine, the engine for 'raising water by means of fire', to bring out the quantitative relations between heat and work.

The early Newcomen steam-engine almost failed because the amount of work done by it hardly paid for the coals used, which were very expensive away from mines or tidewater. A horse could do it cheaper. Watt himself, in order to assess what he was to charge for the use of his engines, had measured the work a horse could do in foot-pounds per minute and had expressed the power of the engines in his new universal *horsepower*. The ingenious method by which the firm of Boulton and Watt managed to sell their engines was to offer to instal and service them free of cost; and to charge a royalty of one-third the saving of cost of either fuel or fodder over a Newcomen engine or horse-gin.[5.41]

The converse action of turning horse-power into heat was first demonstrated by Count Rumford in Munich in 1798 (p. 538). Always interested in heat, particularly in relation to its economical use, he had first noticed and then measured the heat given off in boring cannon. By showing that an indefinite amount of heat could be produced from a limited amount of matter he had effectively disproved the material theory of heat, but this was not enough to establish the alternative theory.

CARNOT: THE REVERSIBLE HEAT-ENGINE

The transformation of heat in the boiler of an engine to power in the flywheel, though amply made use of, could not for a long time be brought into the orbit of exact science.[5.3] Each engine had its own conversion factor of coal burnt into work done, and this factor seemed to decrease as engines improved. No limit to efficiency seemed to be in sight, yet such a limit must exist or perpetual motion would be possible. It was such considerations that led Sadi Carnot, one of the great unrecognized geniuses of the nineteenth century, to his *Réflexions sur la Puissance Motrice du Feu* (1824). Sadi Carnot (1796–1832) was the son of Lazare Carnot, the 'organizer of victory' of the French Revolution. He was trained as an engineer at the new École Polytechnique, and was one of the first to apply mathematical physical principles to the operation of the new machinery.

Carnot conceived the steam-engine as a kind of mill in which *caloric* at a high temperature flowed through the engine and left it in the condenser at a low temperature; provided none was lost in the process the maximum possible work would be done. The test for this was the reversibility of the engine, which, acting as what we now call a *heat pump*, could use the same energy in reverse to raise the same amount of caloric from the low to the high temperature. He showed that even under this optimal condition of *reversibility* only a fraction of the heat put in could be changed into useful work. In other words, work could be done only by the transfer of heat between different temperatures, equivalent to what was later called the *second law of thermodynamics*.

Carnot had gone further than this and had seen that some of the heat was actually transformed into work in the engine, and even found out how much. Before he could publish this knowledge, however, he died of cholera, and his great discovery of the mechanical equivalent of heat remained buried in his notebooks for fifty years. Meanwhile, his published work was also nearly forgotten till rescued by Clapeyron in 1832. Later, however, it was to form the fundamental basis of the new science of thermodynamics. The full elucidation of the relations between heat and work had yet to wait nearly another quarter of a century.[5.3] By that time it was long overdue.

THE CONSERVATION OF ENERGY : MAYER, JOULE, HELMHOLTZ

The first to estimate the *mechanical equivalent of heat* was Robert Mayer (1814–87), a ship's doctor, in 1842. Soon afterwards it was also proposed by Joule (1818–89), a scientific amateur and the son of a wealthy brewer, and by von Helmholtz (1821–94), a physiologist and physicist; and substantially the same idea, though not so clearly expressed, seems to have occurred independently to at least five other physicists or engineers. The approaches of the three principal discoverers were characteristically different. Mayer was led to the conception by general philosophical considerations of a cosmical kind. He was struck by the analogy between the *vis viva* (energy) gained by bodies falling under gravity and the heat given off by compressed gases. Joule was led to the idea first by experiments aimed at finding out how far the new electric motor could become a practical source of power. In showing that it could not, because all the power came from the burning up of the very expensive zinc in the battery that drove it, he was led to consider the quantitative equivalence of work and heat. This he communicated to the British Association at Cork in 1843, but received scant attention. The Royal Society refused to publish his paper in full, and Joule had to batter his way to recognition by ever more accurate experiments.[5.3]

Helmholtz in 1847, by an attempt to generalize the Newtonian conception of motion to that of a large number of bodies acting under mutual attraction, showed that the sum of force and tension, what we would now call kinetic and potential energy, remained the same. This is the principle of the *conservation of energy* in its most formal sense, but

177. The apparatus that James Joule (1819–89) designed for his famous experiment that gave precise evidence for the relationship between mechanical work and the heat generated. The mechanical work done is measured by the movements of the weights (*w*) and heat is generated by the movement of the paddles in the canister of water at A. From a contemporary textbook.

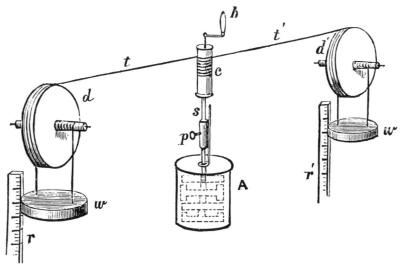

it was important in that it reconciled the new doctrines of heat with the older ones of mechanics, a process that was to be largely completed by William Thomson, later Lord Kelvin, a friend of both Joule and Helmholtz, in his paper *The Dynamical Equivalent of Heat* (1851).*

However varied the approach, all the discoverers were influenced, and more directly than indirectly, by the atmosphere of the age of steam,[5.3] and particularly by the locomotive. As Mayer remarked:

It is in the locomotive that heat is distilled out of the boiler, turned into mechanical work in the moving wheels, and condensed again to heat in the axles, tyres and rails.

The principle of the conservation of energy, of which mechanical work, electricity, and heat were only different forms, was the greatest physical discovery of the middle of the nineteenth century. It brought many sciences together and it fitted very well into the trend of the times. Energy became the universal currency of physics – the gold standard, as it were, of changes in the universe (p. 174). What had been established was a fixed rate of exchange between the different energy currencies: between the calories of heat, the foot-pounds of work, and the kilowatt-hours of electricity. The whole of human activity – industry, transport, lighting, ultimately food and life itself – was seen to depend on this one common term: *energy*.

THE AVAILABILITY OF ENERGY

In the latter part of the century, however, the doctrine of energy which had seemed so optimistic was seriously modified by the realization that the second law of thermodynamics showed that it was not so much the quantity of energy in the universe but its availability that mattered, and that this was always decreasing. In the molecular terms of Maxwell, any system starting with fast (hot) molecules and slow (cold) molecules would end up with most of the molecules moving at intermediate speeds (tepid) or, in the expression of Gibbs (1839–1903), the muddled-upness (entropy) of a system always tended to increase.

If the universe was treated as a whole, it appeared inevitable that the sources of heat would gradually wear themselves out into a universal tepidness, the so-called 'heat death' of the universe. Kelvin, the great propagator of this idea, seemed almost to rejoice in this prospect of a universal mediocrity. Coming nearer home, he was able to prove that the sun could not have been shining indefinitely, and thus that the earth could not have existed for more than a few hundred million years. This was far less time than the geologists required to explain evolution, but the authority of the physicists carried the day. They were wrong, for this prediction, like many others, was doomed to be violently upset by the discovery of new sources of power in the atom of an altogether greater order of magnitude. It is only fair to Kelvin to point out that he guarded himself against this by qualifying his predictions: 'Unless sources now unknown to us are prepared in the great storehouse of creation.'[5.125]

THE PHILOSOPHY OF ENERGY:
MACH, OSTWALD, AND THE NEW POSITIVISM

It was in this period, too, that the knowledge of thermodynamics began to reach into chemistry and even into biology, thanks largely to the works of le Chatelier (1850–1936) and Gibbs (1839–1903).[5.6] It seemed for a while as if the whole of natural phenomena could be explained in terms of simple observables of mechanical energy and heat, and this, in the hands of philosophers like Mach and chemists like Ostwald, seemed to promise an escape from the awkward materialism and radicalism of the atomic theory.

A new positivism appeared which stated that matter and physical hypotheses such as atoms were no longer necessary, and that the whole of science could be deduced directly from elementary observations. The kinetic theory of heat, evolved by Maxwell in 1866 and implying the existence of atoms, was in contradiction to this tendency. Maxwell's atoms, however, were entirely hypothetical and new evidence was needed before they could be accepted as measurable and countable material objects.

9.2 Engineering and Metallurgy

One dominant feature of the eighteenth and nineteenth centuries was the triumph of the machine. Here, however, the part of science is still a relatively minor one. For both in engineering and metallurgy the technical element, based on the tradition of hand work, and the economic

178. James Nasmyth's steam hammer in operation at Thomas Firth's steel works, Sheffield. *c.* 1890.

element, based on profitability, were predominant. Nevertheless the scientific element was always active and grew steadily in importance, preparing the way for the lead it was to take in the twentieth century.

The history of engineering in its great creative phase of the eighteenth and nineteenth centuries marks a continuous interplay between the growing requirements of commerce and industry, and the new means of operation – machinery, engines, materials, which created new possibilities of profitable use. It was the need for more yarn and more cloth that led to the first introduction of textile machinery; the need for more coal, to the first steam-engines; the need for cheap transport of ever more abundant goods led to improvement in ports, canals, roads, bridges, and to the radical innovation of railways (p. 547). However, no sooner was some new mechanism or material developed to meet these needs than new ventures and extensions to other hitherto impossible or unthought-of uses became possible. Thus the steam-engine, first developed for pumping, was adapted next to blowing furnaces and hammering iron, and then to supplant the water-wheel in driving machinery. Later still, mounted on a boat or a wagon, it became automotive and gave birth to the steamship and railway. In a similar way, cheap iron and cheap steel, called into being by specific needs of machine construction, provoked a revolution in the construction of further machines, vehicles, ships, and buildings.

THE ENGINEERS

At every stage in the development of machinery and metals the handicraftsmen were busy trying out new devices and absorbing as much science as they could turn to use; and the scientists were perforce learning the trades so that they could understand the principles underlying them. The process is one we can study through the medium of the biographies of the engineers of the great period from 1750 to 1850, and here we are fortunately well supplied through the work of the great historian of industrial Britain, Samuel Smiles,[5.113-15] and also that of a new generation of more scholarly historians such as Dickinson[5.40-42] and other members of the Newcomen Society. In Britain, for long the centre of the Industrial Revolution, the engineers for the most part began as simple workmen, skilful and ambitious but usually illiterate or self-taught. They were either millwrights like Bramah, mechanics like Murdock and George Stephenson, or smiths like Newcomen and Maudslay. Hardly separable from them, except for their closer connexion with science, were instrument-makers like Smeaton and Watt, artists like Nasmyth (1808–90), or mining engineers like Trevithick. In France, where the workshop played a smaller part and the State and the military schools a larger one, the school-trained engineers predominated: men like Jars, Monge, Poncelet, Fourneyron, Sadi Carnot, and Marc Brunel (1769–1849), a gift of French engineering to Britain. In the later period, after 1850, scientific predominance is more marked, and with it the new importance of Germany in major developments; Britain has only Parsons to balance against Germany's Siemens family, Otto, and Diesel.

The major trends of the whole period of the Industrial Revolution were in the invention of ever more ingenious mechanisms and in the steadily improved performance of machines and structures. Except where new physical principles were involved for the first time, as in new heat-engines and electrical machines, neither made much demand on science. The design of mechanisms, mostly in imitation of a human workman's activities, involved a practical kind of mechanical mathematics too complicated to be learned at school, and stemming rather from the traditional ingenuities of the clock-maker and locksmith. To be successful, however, this had to be matched with a shrewd appreciation of the needs of the industry of the time and a knowledge of where labour saving was likely to be possible and profitable. As these kinds of judgement rarely ran together, the exploiter of inventions – like Arkwright (p. 521), the great promoter of the Industrial Revolution in the cotton trade – usually tended to supplant the mere inventor, who was as likely as not to be ruined; but the machines got built. From 1750 onwards the combination of inventor–exploiter was invincible. Ingenious mechanical substitutes for human hands spread from the textile industries to hundreds of others, both in the manufacture of consumption goods and in the metal and machine industry itself. They even invaded the oldest traditional occupations of agriculture and food processing, especially in America, where, despite slavery, good land was more abundant than were people to work it. Varied as they were and great as was their effect on the growth of civilization, the mechanisms of the eighteenth and nineteenth centuries were combinations of old principles, rather than applications of new as were to be those of the twentieth, and consequently they neither owed much, nor gave much, to science.

EFFICIENCY AND UTILITY:
THE TURBINE AND THE INTERNAL-COMBUSTION ENGINE

The improvement of the *performance* of machinery and engines, almost exclusively steam-engines, was to be the task of successive generations of engineers. Through most of the period it was largely a question of adapting the engine to its various uses, and steadily increasing its power

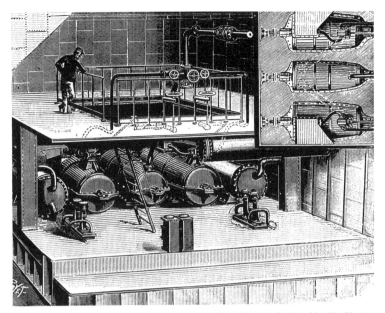

179. Steam turbines fitted into the Dover packet boat *Queen*, designed by Sir Charles Parsons (1854–1931). From *La Nature*, Paris, 1904.

yield per unit of weight of fuel or prime cost, by detailed improvements and better design. In the latter part of the nineteenth century Carnot's ideas, and the thermodynamics built on them, gradually permeated the engineering world; but these ideas were more effective in the revolutionary sense of leading to the turbine, the internal-combustion engine, and the refrigerator rather than in improving the old reciprocating engine.

The new developments were to split the world of power generation into two more manageable and adaptable halves. The *internal-combustion engine* was to lead to the light power unit, to the motor vehicle, and later to the aeroplane; the steam *turbine* to giant ship propulsion and to the generation of distributable electrical energy. Though products of the nineteenth century, they were to find their field of effectiveness only in the twentieth (pp. 806 ff.).

ENGINEERING CONSTRUCTION: THE MACHINE TOOL

The opportunities for profit arising from the use of machinery called into being the machine-building industry; and this in turn was to create a revolution in handicraft, taking the mechanical process one stage further and using machines to make machines. Of these the first and more important were Maudslay's slide-rest and screw-cutting lathes.*[5.12] The debt of this revolution to science was small, and limited to the control of eye and fit judgements by a more rigorous application of geometry, such as Maudslay's plane and micrometer and Whitworth's standard screws. Here the old tradition of the millwright and clockmaker blended continuously into that of the new *mechanical engineer*. The condition that made this possible was the availability of metals – first iron and then steel – capable of taking the new precise shapes, and that of the mechanical power to work them. Only towards the middle of the nine-teenth century did the tasks of engineering begin to get beyond the scale

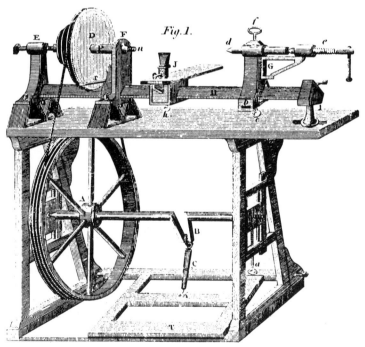

180. The lathe with a sliding tool rest (J) and adjustable headstock (G) designed by Henry Maudslay (1771–1831). This lathe was a great advance and permitted greater precision in turning. From Alexander Jamieson, *A Dictionary of Mechanical Science, Arts, Manufacture, and Miscellaneous Knowledge*, London, 1833.

of resources of the Ancients. Nasmyth's steam-hammer broke once and for all the traditions of Vulcan's forge, and the building of machines became a machine-sized and no longer a man-sized job.[5.90]

Though the actual production of precisely finished metal parts owed little to science, depending as it did on the smooth performance of machines, it was to be the way in which mechanical engineering could itself become scientific. The most elaborate mathematical applications of Newtonian mechanics in the eighteenth century were of little use to practical engineers, because machinery could not be made accurately enough except by the highest craftsmanship and for quite exceptional machines like clocks. Even for the vital needs of war, guns could not be made with sufficiently smooth and uniform bore to enable any serious use to be made of the well-established theories of ballistics.[4.82] With precision metal-cutting all this changed, and the performance of mechanical devices could be calculated from the drawing-board with some chance of predicting performance in advance. It was also to open the way to the use of interchangeable parts, and thus to the mass-production methods of the twentieth century. The first anticipations of this were Eli Whitney's (1765–1825) musket factory in 1800 and the factory for naval stores which Sir Samuel Bentham, Jeremy Bentham's brother, set up in Russia in 1784, and which afterwards led to the British Admiralty block factory in which the machines were made by Maudslay. Both were, significantly, technology for war.

THE METAL REVOLUTIONS

The demands for new machinery, particularly the heavy machinery for mines and later that for railways, ships, and buildings, not to mention the ever-recurring military claims, could only have been satisfied by an ever-increasing flow of metal, and of metal of better and better quality. The ready availability of iron and steel, and the revolution in metallurgical technique that this entailed, were factors in the Industrial Revolution of comparable importance to the invention of textile machinery and to the steam-engine. Here again, as in the case of machine-building, the metallurgical revolution owed much to practical men and little enough to science until the crucial stage in large-scale steel production towards the end of the nineteenth century.

The metallurgy of iron and steel had been practised as a craft for at least 3,000 years. The skill of medieval smiths, of both the East and West, could hardly be improved on. But their products, carefully handmade, were costly and the quantity available was limited to supplying the fairly static demands for axes, horseshoes, ploughshares, arms, and armour. The new demands for artillery for the wars of the sixteenth century strained production in western Europe to the limit, even after the radical invention of cast iron (p. 412). For basic iron production still depended on wood charcoal, and the progressive exhaustion of supplies drove the iron industry into the forests of Sweden, Russia and America.

THE AGE OF IRON

It was this limitation, in the face of the ever-increasing demands of growing commerce and industry, that forced the revolutionary transition from wood charcoal to *charked* pit coal or *coke* in the early eighteenth century, and this completely established the dominance of the coalfields over the forests, for coal as a domestic and industrial fuel had already supplanted wood. Though the possibility of using coal for making iron had long been appreciated, as we have seen (p. 414), actually success turned on the solving of numerous physical and chemical problems quite beyond the science of the day. They had to be solved in practice together with the overriding problem of selling at a profit. The failure of the first projectors lay largely, as in Sturtevant's case, with over-ambitious financing and attempts to enforce monopolies (pp. 414 f.).

Only the perseverance and the probity of the Quaker family of the Darbys of Coalbrookdale [5.98] overcame all these obstacles, and by the middle of the eighteenth century had inaugurated the era of cheap cast iron. The price of pig iron in 1728 was £12 a ton, by 1802 it had fallen to £6.[5.2] But cast iron had its limits. True, rails, pillars, bridges, wheels, engine cylinders could be made of it, but not tools or the working parts of engines. Wherever tension or toughness was needed, wrought iron had to be used, and steel if hardness and springiness were needed as well. Partial solutions to the production of these were found with Huntsman's crucible steel, 1740, and Cort's puddling and rolling process, 1784, both inventions involving much intelligence but owing nothing to official science. Earlier in the eighteenth century Réaumur's work on *L'Art de Convertir le Fer Forgé en Acier* (1722) revealed both the limitations and the possibilities of the science of the time. Réaumur had been able by careful experiment to solve the mystery of the steel-makers, a secret guarded from the time of the Chalybes (p. 148), that steel is iron containing not too much and not too little carbon. He found he could make it by melting cast iron and wrought iron together. He published his results, and in doing so penned one of the noblest defences of freedom of scientific publication,[4.33.151] but no one took any advantage of them. Either the ironmasters could not read or they found Réaumur's recipes impracticable.

Throughout the late eighteenth and early nineteenth centuries the production of iron went on at full blast, with that of steel lagging far behind. Improvements were all in the direction of speeding up the process by the use of a compressed and then a hot blast introduced by Neilson (1792–1865), a gas-works chemist.[5.4] They involved little more than the use of the new mechanical powers to transform an age-old process.

THE AGE OF STEEL:
BESSEMER, SIEMENS, GILCHRIST THOMAS

The decisive break came with the radical innovations of Bessemer in discovering a way of making cast steel on a large scale. In his converter, air, blown through melted pig iron, burns away the carbon, producing enough heat to keep the resulting steel melted. This may be called a semi-scientific result, for though it lacked a theoretical foundation it was

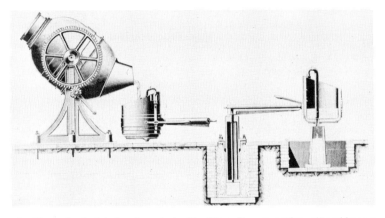

181. The method of steel-making devised by Henry Bessemer (1813–98) and known as the 'Bessemer process' revolutionized the whole process of manufacture, and is still the main method used. Previously laborious and costly methods needing an additional heating process in a furnace or by a complete re-smelting in closed crucibles had been necessary; Bessemer provided a less expensive and eminently practicable solution. He adopted the method of blowing hot air through molten pig iron whereby the temperature was raised, the carbon in the pig iron acting as fuel, and as a result some of the carbon combined with the iron to produce the steel. His ingenuity lay in his design of a tilting converter (left) which was charged with molten metal when in the horizontal position, then tilted into the vertical to receive hot air blasts fed through the bottom, and which was tilted again to discharge the molten steel in a ladle (as shown in the illustration) and from thence into the mould (discharge of ladle shown on the right of the picture). From *Sir Henry Bessemer, F.R.S., An Autobiography*, London, 1905.

arrived at by experimentation. Bessemer was not a scientist but a typical inventor, who knew just enough and not too much science and had a little experience of metals, but not in the iron industry.[5.3] It is notable that neither the ironmasters nor the professors of metallurgy ever proposed any such crazy processes; they knew enough to be sure that they would not work.

Soon after the appearance of Bessemer steel in 1856 an older process took on a new lease of life through the application to the open-hearth or reverberatory furnace of Siemens' principle of heat regeneration, by which the temperature can be raised by using the spent hot gases to heat the incoming air. In this way large charges of steel could be melted and Réaumur's process could be used starting from pig, scrap, and ore. From 1867 on the open-hearth became a serious rival to the Bessemer converter.

Both processes had one serious limitation: they were usable only with relatively pure iron ores (which were not of widespread occurrence) such as those of Sweden, Spain, and Lake Superior. Before they could be used for the more abundant sedimentary ores of Cleveland and Lorraine one final improvement had to be made: the introduction of the basic lining to absorb the deleterious phosphorus. This was the discovery of Gilchrist Thomas in 1879, and is significant not so much because of the magnitude of its consequences, but because it was scientific through and through.[5.3; 5.124] Though Thomas had to earn his living as a police-court clerk in Stepney, he was a master of metallurgical theory; he understood exactly what he was trying to do and the experiments he made in a London cellar could be translated successfully within three years to full-scale production. His work is a portent of the industrial research of the next century.

These three processes together inaugurated the age of steel, first rapidly completing the displacement of wood as a structural material in engineering, and then that of cast iron for rails, ships, and guns. Cheap steel was the basis on which the imperialism of the late nineteenth century was to be built, with its emphasis on ocean commerce, the exploitation of tropical colonies with railway and port developments and its ever more costly preparations for naval and land warfare.

9.3 Electricity and Magnetism

The first new science to arise after the end of the Newtonian period was electricity, in part because it was almost the only aspect of physical science to which Newton himself had not devoted his attention, and accordingly where his great prestige did not frighten off lesser investigators. Electricity had had a long and legendary past. From the earliest times we know of, men had treasured amber and probably noticed its power, when rubbed, of attracting small bodies. It was natural to make the analogy between this and the much stronger power of attraction of the magnet; natural, too, to assimilate both of them into the general magical thinking of ancient times. The doctrine of affinities and attractions, the whole idea of *virtue* residing in a special kind of substance and being evoked by appropriate treatment, was exemplified in amber and even more so in the magnet, because of its magical property of transferring its virtue to other objects by touching them.

The science of magnetism, however, only began when this virtue could be used to good purpose, as in the mariner's compass (p. 317). We have already discussed some of the steps by which the study of the compass led through Peter the Pilgrim and Robert Norman to Gilbert and the beginning of the scientific study of magnetism (pp. 434 f.).

Gilbert's *De Magnete* was not only concerned with magnets; it included a generalization of the attractive principle to cover that of amber and the invention of the first electrical instrument, the balanced pointer or versorium, the later descendants of which, electroscopes and galvanometers, were to give so many pointer readings to science.

EARLY ELECTRICITY: EFFECTS OF FRICTION

Although, as we have seen, Gilbert's magnetism was to be an inspiration to the formation of a theory of gravitation, his electrical experiments were hardly developed beyond the point at which he left them throughout the whole of the great experimental period of the seventeenth century. In its early stages it did not seem to promise any profitable application. It was a philosophic toy and as such lay a little outside the interests of the time, which were turned so largely to mechanics and the vacuum. Nevertheless, some experiments were made in connexion with the vacuum which provided the link with the great developments that were to come later. Von Guericke, the inventor of the vacuum pump, in about 1665, developed the rotating globe or sphere from which, by friction, he drew sparks. This was to be the type of the electrical machines of the next hundred years; but for him it was a model to illustrate his cosmological theories. Picard (1620–82) noticed in 1675 that a barometer shaken in the dark gave a green light – the mercurial phosphorus. This roused the interest of Hauksbee (d. *c.* 1713), Newton's assistant, at the beginning of the eighteenth century. He showed that friction at the same time as generating electricity could produce luminous effects in a vacuum – the forerunner of all our fluorescent lighting – but he made no advance in understanding how they occurred.

GRAY : CONDUCTORS AND NON-CONDUCTORS

Another follower of Newton, Stephen Gray (*c.* 1666–1736),[4,50] pursued similar experiments which led him in 1729 to an illuminating discovery of the transmission of electricity. Almost by accident to start with, but then logically step by step, he was led to the idea that electricity, produced by rubbing a glass tube, could be communicative over large distances. His first observation was that the corks he had put in the ends of his tube attracted small pieces of paper or metal. Next he thought of sticks at the end of the corks, then knobs at the end of the sticks, then balls attached to strings, all of which attracted just as well. Finally he led the electricity out of his room by packthread on silk loops right round the garden and produced what was effectively the first electric telegraph. The fundamental discovery he had made was that electricity was something that could flow from one place to another without any appearance of movement of matter – that it was weightless, an *imponderable fluid*. Electricity could be held in the bodies like glass or silk in which it was generated. These he called the *electrics* – what we now call non-conductors or dielectrics – and electricity could not flow through them. On the other hand electricity flowed through metals or damp string and could not be generated in them. They were the *non-electrics* or *conductors*.

DUFAY : TWO KINDS OF ELECTRICITY

The news of these experiments, so simple and interesting, soon got round and began to make electricity a fashionable and amusing subject which a few amateurs here and there followed up. Dufay in France found in 1733 that there were two kinds of electricity, vitreous and resinous, according to whether glass or amber was rubbed. Many people began to build electrical machines to try all kinds of experiments and even to exhibit them for money.

182. The invention of the Leyden jar in the mid eighteenth century allowed electric charges to be stored and higher voltages to be obtained. The jars contained sheet metal inside and the outside was made of glass. Electric charges were stored on the metal, positive charges on one sheet and negative on the other. More significantly, the Leyden jar in due course led to the development of the condenser, later to become a component of electric circuits. A 'battery' of Leyden jars is here shown, chargeable from an electric machine. From Abraham Rees, *The Cyclopaedia; or, Universal Dictionary of Arts, Sciences and Literature*, London, 1820.

THE LEYDEN JAR AND THE ELECTRIC SHOCK

A fairly obvious idea was that of trying to store the electric fluid in bottles. In 1745 von Kleist (d. 1748), a Pomeranian clergyman, attempted to pass electricity into a bottle through a nail. Touching the nail while holding the bottle in his other hand he received what must have been the first artificially produced electric shock. Some months later, and apparently independently, Musschenbroek (1692–1761) reported a similar experiment from Holland. As he was a scientific apparatus maker, with numerous connexions in the learned world, his name is usually associated with what is still called the Leyden jar.

This discovery had a literally explosive effect. Everybody wanted to try the shock and to see it tried on other people. Electricity became the high fashion in courts. The king of France organized the electrification of his whole brigade of guards, who were made to jump in unison by shocks from batteries of Leyden jars.

FRANKLIN: POSITIVE AND NEGATIVE ELECTRICITY

So much was electricity the rage that Franklin (p. 526), in remote Philadelphia, heard of it and sent for some electrical apparatus. With his robust common sense and apparatus of his own devising, he was able to see through the confusions of previous electrical experiments and proposed the explanation, which holds to this day, that there are not two kinds of electricity but one. He imagined it as a kind of immaterial fluid existing in all bodies, undetectable as long as they were saturated with it. If some was added they became positively charged, if some was removed – negatively. The tendency of the electric fluid to reach its true level was the cause of electric attractions and, when strong enough, of sparks and shocks. If we replace the fluid by practically weightless electrons and change the sign of the charge, — for +, for it is a negatively charged body that has an excess of electrons, Franklin's explanation becomes the modern theory of electric charge.

THE LIGHTNING CONDUCTOR

This simplification, together with an explanation of the action of the Leyden jar, were Franklin's serious contributions to electrical theory and immediately established his scientific reputation. But what really impressed the world at large was his appreciating the analogy between the electric spark of the laboratory and the lightning which he snatched from the sky with his kite and showed that it was electricity. From this he, in his practical way, immediately drew the conclusion that it would be possible to prevent the damage due to lightning, particularly heavy in the New World, by the *lightning conductor* which he tried out in 1753. With this invention electrical science became for the first time of practical use. Franklin's patriotic, or rebel, tendencies had a curious by-effect in England, where in 1780 King George III insisted that the lightning conductors at Kew Palace should have round knobs, instead of the sharp points Franklin had proposed, and Sir John Pringle (1707–82), the President of the Royal Society, who could not agree, was induced to resign. A contemporary wit summed up the controversy with this epigram:

> While you, great George, for safety hunt,
> And sharp conductors change for blunt,
>   The nation's out of joint.
> Franklin a wiser course pursues,
> And all your thunder fearless views,
>   By keeping to the point.

COULOMB AND THE LAW OF ATTRACTION

Despite all these advances electricity and magnetism remained mysterious imponderable fluids, and their quantitative study could not begin until some method could be found of measuring them. This was the work of Coulomb (1738–1806) in 1785, undertaken, significantly, with the object of improving the mariner's compass.[5.12] He found a way of suspending the needle on a fine fibre and used it to measure the forces between magnetic poles and later between electrical charges. This is the torsion balance, the prototype of most sensitive electrical instruments of today, which was also independently developed by Michell (1724–93) and used by Cavendish (1731–1810).* With it Coulomb established what had already been surmised for some years, that the forces between magnetic poles as well as those between charges of electricity obeyed the same laws as those of gravity, that is, a force proportional inversely to the square of the distance. The same conclusion had already been seen to follow from the observation, made by Priestley in 1766 and more carefully by Cavendish in 1771, that no charge was to be found *inside* a charged conductor. These experiments enabled the whole apparatus of Newtonian mechanics to be applied to electricity, but with this difference: that in electricity repulsive as well as attractive forces were to be found.

ANIMAL ELECTRICITY: GALVANI

The immediate development of electricity was not, however, to lie along this quantitative line. Once again, as in the case of the Leyden jar, human and animal sensation came in to reinforce and direct the progress of physics. Acute observers had noticed that there was a close similarity between the shocks given by the Leyden jar and those produced by various electric fishes, particularly by the electric ray (or torpedo – the 'putter to sleep'). Cavendish in 1776 had actually made a working model torpedo out of leather, connected to a battery of Leyden jars.[5.83] This led to the concept of animal electricity, and many confused and ineffectual attempts were made to discover it until in 1780 Galvani (1737–98), professor of anatomy at Bologna, happened to make experiments in which animal preparations were mixed up with electrical apparatus. He noticed that several pairs of frogs' legs contracted whenever there was

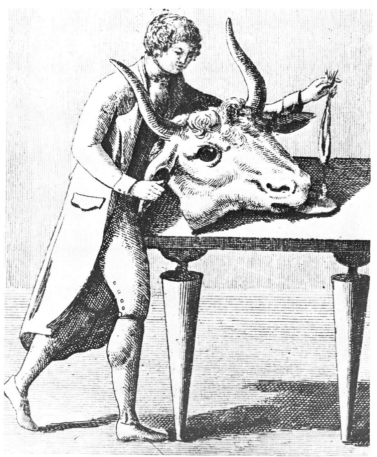

183. The discovery of animal electricity by Luigi Galvani was pursued by others, and not least by his nephew Giovanni Aldini (1762–1834), who lectured on the medical effects of electricity at Bologna at Guy's and St Thomas's Hospitals in London, while his book *Essai Théorique et Experimental sur le Galvanisme*, Paris, 1804, contained a discussion of these.

a spark. It was six years, however, before he observed that it was not really necessary to have the electrical apparatus, that the frogs' legs would contract if two different metals in contact were applied to the nerve and the muscle.

THE ELECTRIC CURRENT AND THE BATTERY: VOLTA

Galvani had in fact discovered current electricity but he did not recognize it. His interest in the physiology of nerves caused him to see his experiments rather as proof of animal electricity. It required the more logical mind of his compatriot Alessandro Volta (1745–1827), professor of physics at Pavia, to understand what he had done. In 1795 Volta showed how to produce electricity without any animal at all by simply putting two different pieces of metal together, with liquid or a damp cloth between them, and he thus produced the first *electrical current battery*.[4,123]

The progress of electricity in the closing decades of the eighteenth century is a clear example of the converging effects of all the sciences, and the particular stimulus given in that time of revolution to everything at once novel and useful. On account of its physiological effects electricity interested doctors and quacks looking for new methods of treatment. Among them was Dr John Graham, whose Temple of Health was presided over by Miss Emma Lyons, afterwards Lady Hamilton. At the same time, and also partly through the doctors, electricity was drawn into the service of the then culminating pneumatic revolution in chemistry (p. 620). In 1800 Dr Carlisle (1748–1840), a London surgeon, and his friend William Nicholson (1753–1816), an engineer, commercial traveller, and scientific publisher, used the newly invented battery to decompose water into its constituents – oxygen and hydrogen (p. 623). Thus they settled a crucial problem in chemistry and started the new sub-science of electrochemistry.

Galvanic batteries soon were as much a necessity for a well-equipped laboratory as batteries of Leyden jars had been fifty years before. But at first they were even more expensive, and only the wealthiest could build large ones. Thus it fell to Davy in 1802 to produce the new metals, sodium and potassium, by the use of the world's largest battery at the Royal Institution. These experiments brought electricity out of its isolation, as a set of peculiar phenomena, and linked it with the general body of science. It was beginning to show a promise of use as well as interest. The promise could not, however, be fulfilled for some decades, until a connexion had been found between electricity and magnetism.

Except for the discovery that the electricity from the Galvanic cell and that from the frictional machine were of the same kind, though vastly

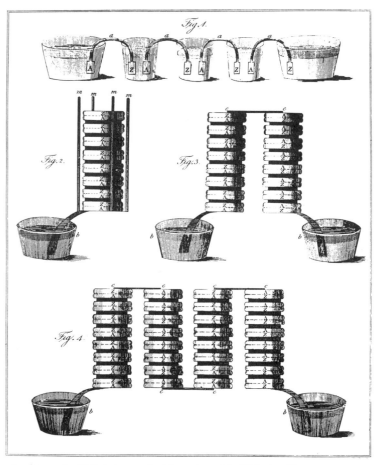

184. Alessandro Volta's (1745–1827) invention of the 'Voltaic cell' was a fundamental discovery. In practice it put a source of continuous electrical current into the hands of the experimenter and the engineer. From Volta's letter in the Royal Society's *Philosophical Transactions*, 1800.

different in quantity and intensity, the nature of the electric current was to remain shrouded in mystery for another twenty years. Currents from batteries were variable and unpredictable and it was impossible to subject them to measurement until an entirely different effect of the current was discovered.

The multiple analogies between electricity and magnetism made physicists think that there must be some connexion between them, but it was one very difficult to find. It was not until 1820 that through another accident at the lecture table, Oersted (1757–1851) in Copenhagen found that the electric current deflected a compass needle. He thus joined together, once and for all, the sciences of electricity and magnetism. One immediate consequence was the invention by Sturgeon (1743–1850) in 1823 of the electromagnet and its improvement by Henry (1799–1878) in 1831.[5,6] At one remove it led to that of the electric telegraph and the electric motor.

The deflection of the compass by the electric current also had enormous theoretical importance. In the hands of Ampère (1775–1836), Gauss (1777–1855), and Ohm (1787–1854) it led to the understanding of the magnetic fields produced by currents and of the way these flowed through conductors. Current electricity could now become a quantitative science and take over all the mathematical apparatus of mechanics. Nevertheless, in one important and puzzling respect the new laws differed from those of Newton. All the forces between bodies that he considered, acted along the line joining their centres; but here a magnetic pole was urged to move *at right-angles* to the line joining it to the current-carrying wire. This was the first break from the simple *scalar* field theory, and opened the way to a more inclusive *vector* theory where direction as well as distance counted. It was these physical discoveries that were to give a new impetus to mathematics and to wean it from the now sterile adherence to the Newtonian tradition.

ACCIDENTAL DISCOVERIES?

It is interesting to reflect on the sequence of apparently accidental discoveries that led to this stage of knowledge. At first sight it seems to reinforce the idea that science is entirely unpredictable and depends entirely on purely chance discoveries. Actually, now that we know the character of some of the relations between different aspects of Nature, we can see that it must have been extremely difficult in the long run not to have hit upon them in one way, if not in another. Oersted, inspired by the unitary ideas of *Naturphilosophie*, had certainly been looking for the connexion between electricity and magnetism for thirteen years, but his actual discovery was not the result of any deliberate planning. In this case, as there were so many people playing with electric currents and compass needles at that time, someone could hardly fail sooner or later to notice their interaction. Many probably did and thought no more of it. The difficulty in science is often not so much how to make the discovery but to know that one has made it. In all experiments there are a number of effects, produced by all kinds of extraneous causes, which are not in the least significant, and it requires a certain degree of intelligence or intuition to see which of them really mean anything. This is particularly so when there is nothing in existing theory to make one expect such events to happen, and even more so when, as is often the case, there appear to be perfectly good reasons for not expecting them. Sooner or later, however, if enough people concentrate on the field, someone will be found sufficiently observant, sufficiently broadminded, and sufficiently critical or ignorant of orthodox theories to make the discovery (pp. 748 f.).[4.12]

MICHAEL FARADAY : ELECTROMAGNETIC INDUCTION

Before the full interaction of electricity and magnetism could be understood, still one more decisive step had to be taken. It had been shown how electric currents produced magnetism; it remained to show how magnetism could produce electric currents. This discovery, though it had to wait for another ten years, was not, like Oersted's, accidental. It was the result of a deliberately planned research by Faraday. In 1831, in his fortieth year, and free from the restrictions that the somewhat jealous Davy put on his work,[5.5] Faraday showed that the relation between magnetism and electricity was dynamic and not static – that a magnet had to be moved near an electric conductor for the current to arise. This most crucial observation showed that not only was magnetism equivalent to electricity in motion but also, conversely, electricity was magnetism in motion. Thus both sets of phenomena could only be discussed in the new joint science of *electromagnetism*.

Faraday's discovery was also of much greater practical importance than Oersted's because it meant that it was possible to generate electric currents by mechanical action, and conversely that it was possible to operate machinery by electric currents. In essence the whole of the heavy electrical industry was in Faraday's discovery, but it took the greater part of fifty years before full advantage could be drawn from it (pp. 612 f.) Faraday himself had little inclination to move in the

185. A page from the experimental notebook of Michael Faraday (1791–1867), showing a sketch of his experiment to prove that magnetism can induce an electric current in a separate coil of wire: a discovery from which the transformer was developed.

Aug 29th 1831.

Expts on the production of Electricity from Magnetism &c

Have had an iron ring made (soft iron) iron round ⅞ inches
thick & ring 6 inches in external diameter. Wound many
coils of copper wire round one half the ring but separated
by twine & calico - there were 3 lengths of wire each about 24
feet long and they could be connected as one length or used
as separate lengths. By twine with a trough each was
insulated from the other. Will call this side of the ring
A. On the other side but separated by an
interval was wound wire in two pieces
together amounting to about 60 feet in
length the direction being as with the former
coils this side call B.

Charged a battery of 10 pr plates 6 inches square. Made
the coil on B side one coil and connected its extremities by
a copper wire passing to a distance and put over a magnetic
needle (3 feet from wire ring) then connected the end of one of the
pieces on A side with battery. Immediately a sensible effect on needle
It oscillated & settled at last in original position. On breaking
connection of A side with Battery again a disturbance
of the needle.

Made all the wires on A side one coil and sent current from battery through the whole. Effect on needle much
stronger than before -

The effect on the needle then but a very small part of
that which the wire communicating directly with the battery
would produce

direction of practical application. This was not due to any other-worldliness; Faraday knew enough from experience of the world of business and government to estimate the time and trouble it would take him to bring any of his ideas to the stage of profitable exploitation. He felt he could make better use of his time.[5.28]

He was concerned, as his notebooks show, with a long-range project of discovering the connexions between all the 'forces' that were known to the physics of his time – electricity, magnetism, heat, and light – and by a series of ingenious experiments he was in fact able to succeed in establishing every one of these, and to discover in the process many other effects the full elucidation of which has had to wait till our time.[5.46]

THE ELECTROMAGNETIC FIELD : MAXWELL

Faraday was one of that rare class of physicists who had a visual and almost sensuous understanding of the forces with which he was dealing. His vivid imagination created the picture of electric and magnetic fields, equipped with lines and tubes of force, showing that whenever a tube of magnetic force cut an electric conductor it gave rise to an electric current, and conversely that the movement of electric tubes of force gave rise to magnetic fields. In this sense Faraday's work was comple-mentary to that great mathematical synthesis of Newton, where fields and potentials took the place of attractions between geometrical points. The formal translation of Faraday's qualitative intuitions into precise and quantitative mathematical equations was the work of Clerk Maxwell (1831–79), who summarized in concise form the whole of electromagnetic theory – apart from the apparently wayward effects of electricity on matter such as occurred in electrical discharges and were to lead to the discovery of the electron (p. 732).

ELECTROMAGNETIC WAVES

But Maxwell's equations did more: from their form it was possible to see that they could be fitted to expressions for waves of electromagnetic disturbance which would travel with a speed suggestively near to that of light. The nineteenth century had already witnessed a great reversal in ideas on the nature of light. Newton had determined to his own satis-faction, and for 100 years no one had dared to question his authority, that light consisted of fiery particles travelling at great speed. In 1801 the physician Thomas Young (1773–1829) in England, and the physicist Fresnel (1788–1827) in France, had been forced, by a consideration of the interference and polarization of light, to go back to Huygens' view that it consisted of waves. After a sharp battle with worshippers of

186. James Clerk Maxwell, whose precise formulation in 1873 of electromagnetic theory led to many of the modern developments in physics and in electronic technology. From a steel engraving by G. D. Stodart in *The Life of James Clerk Maxwell* by Lewis Campbell and William Garnett, London, 1882.

Newton, they carried the day, and for 100 years the wave character of light was unchallenged. However, if the fiery particles were no longer needed, a medium was required to carry the waves, even through the vast emptiness of space, and the 'luminiferous ether', which had the incompatible properties of being infinitely rigid and infinitely tenuous at the same time,[4.168] was called into existence to do so, to be 'the nominative of the verb to undulate'. But electricity and magnetism had also long been known to act through empty space. For them, equally intangible *fields* were created. Maxwell showed in effect that one single but still mysterious ether (p. 480) would do for all three. He had achieved a great condensation and simplification in physics from which important consequences were soon to follow.

One was the establishment of a new unity between sciences: the whole of light appearing as an electromagnetic phenomenon. Another was the inference that electromagnetic oscillations ought to give waves in the ether similar to those of light, but with much lower frequencies. Hertz (1857–94) demonstrated these in the laboratory in 1888, and they were later to become the basis of radio-communication.

With Maxwell's equations, electrical theory appeared to be so nearly complete that the future of physics seemed to hold only an extension and perfection of it. Actually, as we shall see in the next chapter, it covered only a small part of the phenomena of electricity, and the corpuscular unit of electricity, the electron, escaped the equations entirely (p. 627).

THE LAGS IN THE APPLICATION OF ELECTRICITY

In order to present a coherent story of the development of electromagnetic theory it has been told as one logical sequence running right through the nineteenth century. But the growth of electricity throughout that period had another and practical side that interacted continuously with the advance of theory.[5.3] From about 1830 onwards electricity began to contribute directly to economic life, first in the form of communication, then for electroplating, for light, and for power, with two new forms of communication – the telephone and wireless – thrown in at the end of the century. Electricity was indeed the first science to create an industry of its own without any dependence on tradition.

The process was nevertheless a slow one, for in spite of the legend of the alert entrepreneur capitalist seizing on new ideas and marketing them ahead of his competitors, there were in fact enormous practical difficulties in introducing anything that required development before it would pay. Both academic scientists and independent inventors were in constant straits to finance these developments. The only way it could be done was to produce anything that would sell quickly, and to finance each new development from the profits of the last. Very few people managed to surmount all the hurdles of an important application. Most were broken or discouraged, and there were innumerable false starts.

In the process of converting the discoveries of the laboratory into the products of a profitable industry, four main stages can be discerned, each concerned with a different practical utilization of the new electrical principles. They were the telegraph, electroplating, arc lighting, and finally the filament lamp. Of these, the first, as it demanded little current, led mainly to the improvement of batteries and receiving instruments, and thus largely to the development of electrical theory.

Electroplating, on the other hand, called for heavy currents and put a premium on the use of some forms of mechanically generated electricity. This led to the first applications of Faraday's principle, but one which employed only permanent magnets (Pixii's machine) and was thus weak and inefficient. Further, the demand of the electroplating industry could never be very extensive.

ARC LIGHT AND DYNAMO

A far greater field was furnished by arc lighting, and the need for efficient generators was established. It was the discovery by Wilde (1833–1919) and Sir William Siemens (1823–83), in 1867, that the current from one machine could be used to excite the field electromagnet of another that led to the first *dynamo*, the energy symbol of a new age. With relatively cheap current available the emphasis turned to finding extensive uses for it, but here the most promising field was in domestic and shop lighting, for which the arc was too brilliant.

The solution to the problem of 'subdivision of the electric light' was found in the incandescent lamp with a filament, first of carbon, then of metal, in its evacuated bulb. The technical problem of making a cheap and durable lamp was considerable, but it was not this that held up progress; incandescent lamps of sorts had been made in Russia by Lodygin (1847–1923) in 1872 and by Swan (1828–1914) in England a little later. For commercial production, the lamps required a greatly improved vacuum pumping system, but, given the incentive, that could have been achieved at any time in the century. The real difficulty was on the distribution and sales side. Edison's decisive contribution was the *power station* of 1881, with its network of mains serving electricity like gas or water.[5.65]

The fifty years' delay between Faraday's discovery and Edison's application was thus due to no scientific or technical lag, but to essentially

187. Henry Wilde's (1833–1919) design of a dynamo, widely adopted in the 1870s. It used separate coils of wire (field coils) for developing the magnetic field in which the armature rotated. From Georges Dary, *Tout par l'Électricité*, Tours, 1883.

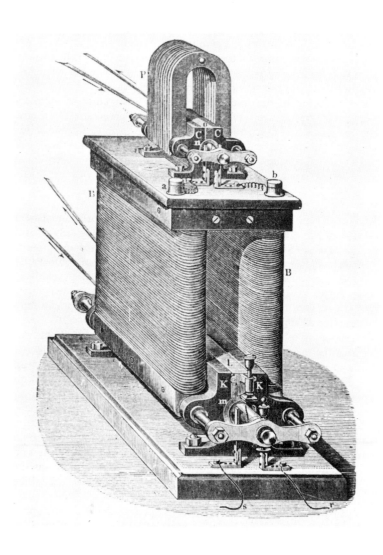

188. Electric light, using an incandescent wire filament sealed inside an evacuated glass bulb to prevent combustion of the wire, came into general use after 1881. Here a mercury pump is being operated for evacuating glass bulbs, c. 1883. From R. Wormell, *Electricity in the Service of Man*, London, 1896, translated from the German of A. R. von Urbanitzky.

economic and social causes.[5.3] No means were available in the mid nineteenth century for an organized exploitation of a scientific idea up to the stage at which it could pay its way. Once that stage was reached there was no holding it. Electric light and power had arrived; they were to expand in the next century at a rate far greater than that of steam.

The role of electricity in power distribution for transport, machine driving, heat, and light, as well as its use in the telegraph and telephone, all depended on an elaboration of the original electromagnetic experiments of Oersted and Faraday, reduced to a mathematical form by a generation of theoretical physicists, culminating in Maxwell. No radically new physical idea had, in fact, been added since 1831. The electrical industry of the nineteenth, and also of the twentieth century, apart from electronic applications, was an ideal example of a purely scientific industry depending on skill and ingenuity in using a limited set of principles for the solution of an ever-increasing range of practical applications.

The story of electricity and magnetism provides the first example in history of the transformation of a purely scientific body of experiments and theories into a large-scale industry. The electrical industry is necessarily scientific through and through. Nevertheless, we find here the most irrefutable example of how at one remove scientific research can turn into engineering practice. There was no need for the men who were going to rig up telegraph systems to have the same scientific calibre as the inventors of the telegraph. This gave rise to the profession of the telegraph engineers who were incorporated in a society in 1871, which in 1889 changed its name to the Institution of Electrical Engineers. Within fifty years electrical engineering had acquired a tradition and code of practice. Problems of design and production, of economy in working and ease of repair had been superimposed on the basic scientific principles of electromagnetic induction. The wheel was in the end to come full circle, and the new profession was for a short time to furnish the livelihood of two young men who were to revolutionize physics, Albert Einstein and P.A.M. Dirac.

THE ELECTRIC DISCHARGE AND THE NEW PHYSICS

The practical triumphs of electrical engineering were not, however, to be the most fruitful ultimate consequence of the pursuit of the sciences of electricity and magnetism. Nor were they to lie in the further pursuit of electromagnetic theory. It was from a totally different set of phenomena – the curious luminous glows that had intrigued the first electrical amateurs of the seventeenth century – that the great new advances were

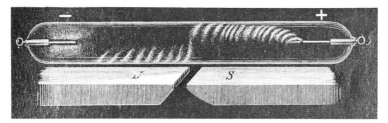

189. The investigation of electrical discharges, glowing colours, and mysterious 'cathode rays' was made at the close of the nineteenth century after Heinrich Geissler's (1814–79) development of a method for welding electrodes on to glass tubes. These studies led to the discovery of electrified particles shown here deflected by magnets, and so to the development of atomic physics. From R. Wormell, *Electricity in the Service of Man* (see plate 188).

to come, leading, as we shall see in Chapter 10, to the discovery of X-rays, the electron, radioactivity, electronic valves, atomic theory, and ultimately to atomic fission. This branch was not an obviously promising one; the phenomena were capricious and almost impossible to reduce to quantitative terms, and no practical applications came to hand to focus interest and lead to intensive research. It was accordingly pursued in a desultory way, and the exciting results which it could produce had to wait till the end of the century to be discovered.

9.4 Chemistry

The central feature of science in the eighteenth and nineteenth centuries was the rise, indeed the establishment, of chemistry as a rational discipline of thought and practice. In the practical sense the science of chemistry was as old as, or older than, any other science; but as already explained (pp. 80, 222, 279) it was not, and could not become, a logical science until very late, since the science of earlier times lacked the essential prerequisites. It was necessary first to wait for the accumulation of a far larger body of experience of the properties and transformations of a greater variety of substances than was available in ancient or Renaissance times. The rapid development of a widespread mining and chemical industry of a non-scientific and essentially technical character was a necessary precondition for the building up of any effective chemical theory. But it also needed some comprehensive ideas which would weld together these diverse experiments and make out of them a coherent picture which could be grasped and used to lead to further discoveries.

THE END OF ALCHEMY

One preliminary requirement for any rational view of chemistry was the removal of the magical beliefs, drawn from classical and even earlier times, which still cluttered up the work of the practical chemist. Of these the most pernicious and difficult to eradicate were the astrological and mystical aspects of alchemy and its preoccupation with the then futile problem of making gold. The first attempt in the seventeenth century to make chemistry rational must, as we have seen (pp. 472 f.), be written off as a failure, though with the work of Boyle, Hooke, and Mayow it came very near to success. The corpuscular philosophy, with its over-rigid mathematical–mechanical models, could not, in fact, be applied to chemistry until its qualitative features had first been more thoroughly elucidated.

THE SEARCH FOR CHEMICAL PRINCIPLES

The line of advance in chemistry was to be, for most of the eighteenth century, a quite different one. Instead of attempting to apply to chemistry rational principles based on mechanical models which could not cope with the enormous diversity of chemical facts, knowledge was to advance by a progressive rationalization of originally magical and animistic ideas. These, though at the outset inevitably vague, had an elasticity which enabled practical chemists to comprehend and set in order in a few verbal generalities all their multifarious operations. Only when this had been done was it possible to apply significantly the physical tests of measurement and calculation. The great advance of the eighteenth century was to narrow down chemical problems to one central one, the problem of *combustion* – the operations of the spirit of fire (p. 280). The question was: What happened to combustible materials when they burnt in air? The obvious answer was that they disappeared in flame and smoke and left ash. This picture, however, was all very well for wood and oil, but was not easy to extend to other substances like metals which mortified or rusted in air. Had all these anything in common and what was the function of the air?

Some answers to these questions had already been provided in the seventeenth century. Jean Rey in 1630[4,127] and Mayow in 1674[4,105] had established the cardinal facts that metals gained weight on heating in air, and that the air itself contained something of a 'nitro aerial spirit' which was concerned in maintaining both fire and the breath of life. But these were isolated forerunners, unable to influence the general stream of chemical thought (pp. 472 f.).

THE DOCTRINE OF PHLOGISTON

Indeed this stream was flowing strongly in the opposite direction, towards the view that all combustibles contained a substance that they lost on burning. This was essentially the sulphur of the Arabs and Paracelsans, but it was given a new lease of life by Becher (1635–82) and his disciple Stahl (1660–1734) by christening it *phlogiston*, the principle of phlox or flame, though the phlogiston theory was only generally accepted by the mid eighteenth century. Bodies containing much phlogiston burnt well; bodies that would not burn were de-phlogisticated. A body with much phlogiston, like coal, could transfer it to a body that had lost it, like iron ore, and by infecting it with phlogiston turn it into shining metallic iron. Even from the start objections were raised to this theory. It was pointed out that phlogiston was not a substance. It was essentially the opposite of a substance; it had no mass. But as we have already seen (p. 602) there was nothing strange in the idea of an imponderable fluid – electricity, magnetism, and heat, all of undoubted reality, were also of that nature. Even when it was established that some bodies actually grew heavier on losing phlogiston, this was put down either to a secondary accretion from the air or to the idea that phlogiston had natural levity.

We are apt, looking at it from the point of view of its immediate successor – the theory of combustion as oxidation – to treat the phlogiston theory as absurd; in fact it was an extremely valuable theory and it co-ordinated a large number of different phenomena in chemistry. It proved a good working basis for the best chemists of the mid eighteenth century, and was firmly adhered to till the end by many of them, including the man whose experiments were to destroy it, Joseph Priestley (pp. 531 f.)

THE LOGIC OF PHLOGISTON

The central concept on which it turned was the universality of the antithetical processes of *phlogistication–dephlogistication*. Thus it brought together processes that were alike, and separated those that were unlike. As its opponents saw it, dephlogistication was not the removal of a metaphysical substance, phlogiston: it was the addition of a material substance, oxygen – *oxidation*; while phlogistication was its removal – *reduction*. It was necessary for the progress of chemistry that the balance should be the test. We can now in the twentieth century afford to reverse this idea again and return to phlogiston as a material, although a very light one; in modern parlance it could be spoken of as electrons.* Those substances that have an excess of easily removable electrons, like hydrogen, metals, or coal, are those which were thought to be rich in phlogiston; those in which there is an exact balance of electrons, like salts and oxides, are dephlogisticated; while those that eagerly absorb electrons, like oxygen, would appear as highly dephlogisticated. The failure of the phlogiston theory was not on account of its internal illogicality, but because as it stood it could never be squared with the material facts. It needed to be turned upside down, phlogistication becoming de-oxidation, and dephlogistication, oxidation. The impetus for this inversion was to come not from traditional chemistry, but from another quarter – the study of gases.

THE PNEUMATIC REVOLUTION:
WILD UNTAMEABLE SPIRITS: VAN HELMONT

By the middle of the eighteenth century distillation was no longer any kind of novelty, and interest shifted to those products of chemical action that could not be recovered in the condenser, the 'wild untameable spirits' of van Helmont (p. 439). Such spirits, ghosts, or gases (*chaoses*), as he called them, were well known in practice, particularly to miners, and were beginning to attract the attention of scientists; they were the treacherous fire-damps and 'inflammable airs' of mines and marshes that could be collected in bladders and burnt. There was as well the deadly *mofette* of caves, the 'afterdamp' that followed explosions in mines, which was also to be found in brewers' vats and asphyxiated the workmen who occasionally fell into them.

HALES AND THE HANDLING OF GASES

It was from the study of these *gases* that the clue to the explanation of chemistry was to be found. The Rev. Stephen Hales (1677–1761) in his *Vegetable Staticks* had already, early in the century, shown how to collect gases over water and to measure their volume. Later Priestley and Cavendish collected them even more effectively over mercury. The next need was to recognize that these gases were not just air, but that there were *qualitative* differences between them. It was then necessary only to bring to bear on gases of different kinds the same quantitative treatment that Boyle had applied to the transformation of bodies.

THE TEST OF BALANCE: THE CONSERVATION OF MATTER

The essential advance was that of extending the idea of weighing chemicals undergoing change to *all* the products of change and not, as in the old assaying, confining interest only to the weight of the original ore. As long as gases entering or leaving the reaction were not weighed or measured, it was clearly impossible to make the books of chemistry balance. That they should do so was first clearly enunciated by Lomonosov (p. 515) in 1748 as the principle of the conservation of matter, but his work was overlooked and it was left for Lavoisier to establish it as a fundamental principle in 1785, curiously enough from a study of the processes of fermentation.

JOSEPH BLACK: FIXED AIR

The first step in the new quantitative pneumatic chemistry was taken by Joseph Black, a Scottish doctor who had his interest roused by Dr Cullen's first chemical lectures in Glasgow. Black wrote his M.D. thesis in 1754 on 'Experiments upon Magnesia Alba, Quicklime, and other Alkaline Substances' in the search for a new and mild remedy for the stone, the most prevalent ailment of the heavy drinkers of the eighteenth century. The House of Commons had voted an award of £5,000 to Joanna Stephens for revealing such a remedy, which was found to consist of calcined snail shells mixed with honey.

Black distinguished and weighed, as loss, the gas given off by carbonates such as limestone or magnesia when heated. He called it 'fixed air' because he could absorb it in lime water and thus reconstitute the original carbonate, with an identical gain in weight. In this way he showed that a gas could be an integral part of a solid body, that it was strictly material and had nothing mystic left about it.

JOSEPH PRIESTLEY AND THE DISCOVERY OF OXYGEN

The next important advance was due to Joseph Priestley (pp. 531 f.). It was in the course of writing a history of electricity, at Franklin's suggestion, that Priestley made certain experiments on electric discharges in air that led him out of the field of physics into that of chemistry. It is characteristic of these early days that radical advances in chemistry were not made by chemists. Chemists knew too much, they had theories that explained everything; it was for the physicists, who knew nothing, to provide fool or common-sense explanations.

Priestley had seized the notion that there was not only one kind of air. He played with as many gases as he could find and made many others. His first success was the preparation of *soda water* containing fixed air in solution. For this he was awarded the highest honour of the Royal Society – the Copley Medal. Though it disappointed the early hope that it would prove a cure for scurvy, the curse of long ocean voyages, it remained on its own merits, the first new commercial product of pneumatic chemistry.

One gas, which he made by heating red oxide of mercury (*mercurius calcinatus per se*), he chose to call 'dephlogisticated air' because it had a greater affinity for phlogiston than ordinary air, that is, things burnt better in it. This was what we now call *oxygen*, and its discovery in 1774 was the culminating point of what may properly be called the pneumatic revolution of chemistry. Scheele in Sweden had also prepared oxygen at about the same time. He was a far better chemist than Priestley, but his interests lay rather in analysis than in the theoretical problems of chemistry, and so his discovery of oxygen did not contribute as much as it should have done to the solution of central problems. Priestley showed that in burning and in breathing alike it was the dephlogisticated air (our oxygen) that was used up. He also showed that in sunlight green plants actually produced oxygen from the fixed air, or carbon dioxide,

190. Joseph Priestley's pneumatic trough allowed him to collect different kinds of gases conveniently, and to discover, among other things, the gas oxygen. From his *Experiments and Observations on Different Kinds of Air*, London, 1774–7.

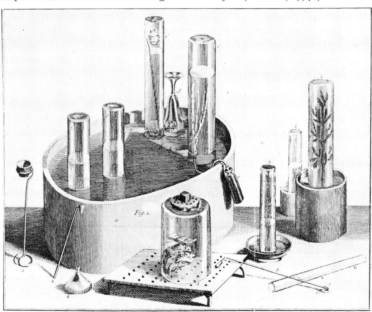

that they absorbed. He had thus solved in principle the essential problem of the carbon cycle: from the atmosphere through plants and animals and back to the atmosphere again. But he did not fully understand the significance of the range of his own discoveries, and it fell to Lavoisier, with his far more logical and well ordered mind, to make up this deficiency.

THE OVERTHROW OF THE PHLOGISTON THEORY

Like Priestley, Lavoisier came to chemistry through physics (pp. 533 f.). Unlike Priestley, however, he did not spread himself in extensive qualitative experiments, but set himself limited and definite tasks of investigating the mechanism of combustion in air, which he saw was crucial to chemical theory. His work was precise, ordered, and *quantitative* throughout. In 1773, already conscious of the importance of the new pneumatic chemistry, and particularly of the fixing of air as a material fact, he formed the project of using it 'to bring about a revolution in physics and chemistry'. Later, hearing of Priestley's discovery of oxygen, he realized its significance at once, and was able to show that it alone was responsible for combustion, which was neither more nor less than the adding of oxygen, originally *le principe oxygine*, the acid-producer – a word he coined for the purpose. This ran absolutely counter to the phlogiston theory, but he did not hesitate a moment in attacking it, reversing all its arguments and putting it, as Marx did with Hegel, on its feet again (pp. 1102 f.).

THE CHEMICAL ELEMENTS

Lavoisier showed that the whole of the previously chaotic phenomena of chemistry could be ordered in a law of combination of elements old and new. To the established list of elements, in the sense of Boyle, not of Aristotle (p. 473) – carbon, sulphur, phosphorus, and all the metals – he added his new oxygen which together with *hydrogen* went to make up the old element water, as well as the other constituent of the air, the lifeless azote or, as we call it, *nitrogen*. According to this new system, chemical compounds were largely of three categories; those of oxygen and a non-metal, which were *acid*; those of oxygen and metals, which were *bases*; and the combination of acids and bases, *salts*. Lavoisier made a clean sweep of all the old time-hallowed chemical nomenclature based on methods of preparation or fancied resemblances: oil of tartar per deliquum, sugar of lead, and so on, and introduced instead the terms we now use – potassium carbonate, lead acetate, etc. This step in itself marked the extension to chemistry of the same rationalizing process that had been applied to physics in the early seventeenth century, and also drew on the simplified nomenclature Linnaeus had introduced in his botanical classification (pp. 636 f.).

Lavoisier himself, however, carried the same process one step further; making use of the rapidly accumulating data on the *quantities* in which various solid substances combined, he extended this to cover the newly found gases, and, thanks to his law of conservation of mass, reduced chemistry to accountancy into which only elements entered. Thus at one stroke he converted chemistry from a set of independent recipes, which had to be known one by one, to a general theory from which it was possible not only to explain the previous phenomena, but also to predict new ones in a quantitative way. Lavoisier was more a legislator for chemistry than a systematic chemist; he seized on essential points, and left to others, such as Berthollet (1748–1821) and Richter (1762–1807), the task of examining the nature of chemical affinity or the precise proportions in which chemical substances actually combine.

THE PRIMACY OF CHEMISTRY

Lavoisier's success in effecting a revolution in chemistry aroused immense enthusiasm. Revolution was in the air, and the new chemistry, now so closely linked to physics, soon attracted to itself some of the most intelligent minds of the time, and helped to secure for France a predominant place in the world of science for nearly half a century.

The interest in chemistry was reflected in industry, and in turn industry supplied chemistry with new substances and new problems. The study of the glass-colouring mineral, manganese, by Scheele had led to the discovery of *chlorine* in 1774. Berthollet in 1784 found its use in bleaching, and McGregor, inspired by his son-in-law Watt, first used it on a large scale in the growing linen industry of Glasgow.[5.4] The other main industrial chemical advances were Roebuck's manufacture of sulphuric acid (1746), which served to replace skim milk sours in bleaching, and soda manufacture from salt instead of expensive kelp and barilla,[5.4] according to the processes of Keir (1735–1820) in 1769 and Leblanc (1742–1806) in 1790. Though Leblanc himself was left to die in poverty, his process was perfected by direct orders of Napoleon, and its success made France independent of supplies of soda from countries controlled by England. All these processes were essential adjuncts to the enormous increase of textile output that was the main growing point of the Industrial Revolution, and was outrunning the limited supplies of vegetable products. Even where, as in these cases, the processes arose from traditional or phlogiston theory, their success and the anticipation of further successes

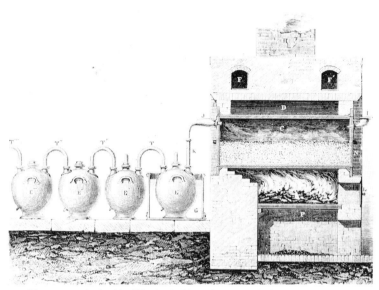

191. The process devised by Nicolas Leblanc (1742–1806), for the manufacture of soda from the decomposition of common salt by sulphuric acid, a far less expensive method than previously employed. At R, in an oven above the fire, sea salt is placed; sulphuric acid is poured through the funnel at I (on right), and hydrogen chloride gas is given off. This is collected in the containers E, E^1, etc., on the left, which contain water. Sulphate of soda remains in the oven. From Louis Figuier, *Les Merveilles de l'Industrie*, Paris, c. 1875.

to come stimulated the study of chemistry, and led to the ready adoption of the new rational doctrines.

THE CHEMISTRY OF EATING AND BREATHING

Lavoisier's other contribution to science was to make quantitative Priestley's qualitative pictures of the chemical nature of the process of life, and he thus became the father of quantitative physiology. By a set of admirably designed and executed experiments he was able to show that a living body behaved in exactly the same way as fire, burning up the materials in the food and liberating the resulting energy as heat. For the first time the general chemical balance sheet of organisms could be established, and the real significance of the mechanisms of breathing and of the circulation of the blood, discovered by Harvey nearly 200 years before, was revealed.

DALTON : THE ATOMIC THEORY

The next crucial step in the understanding of chemistry was taken twenty years later by John Dalton (1766–1844), a Quaker weaver and school teacher of Manchester. He, like Priestley and Lavoisier, was not primarily a chemist but a physicist and meteorologist. He was interested in gases as elastic fluids, and tried to explain their properties on Newtonian principles by the mutual repulsions of the *atoms*. This led him to consider the possible proportions of atoms in different kinds of gases, and thus to see how to explain the laws of combination of elements in multiples of definite weights, which had gradually emerged from the analyses of the new gases such as nitrous oxide, nitric oxide, and nitrogen peroxide, which we write, following Dalton, N_2O, NO, and NO_2.

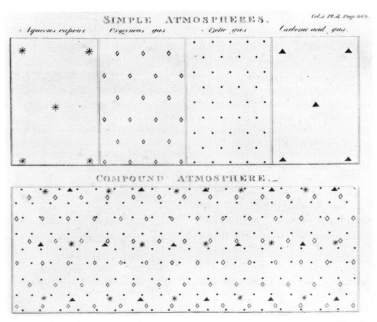

192. Simple and Compound Atmospheres. In 1802 Dalton wrote a paper 'On the Expansion of Elastic Fluids', in which he showed that the pressure of a gas composed of a mixture of other gases is equal to the pressure of all the other gases added together – now known as Dalton's Law of Partial Pressures. This and other works on the physics of the atmosphere paved the way for his atomic theory (to which he made the first reference in 1803), since it led him to consider separate 'ultimate particles' each different for different elements, with each acting on its own, as was the case with his atmospheric gas particles, illustrated here.

These followed simply from the assumption that all chemical compounds were made up atom by atom – the atoms of different kinds arranging themselves in pairs, threes, or fours.

CRYSTALLOGRAPHY: HAÜY

Other regularities, those occurring in crystals, were also about this time pointing to an atomic explanation. Steno in the seventeenth century had shown the invariability of the *angles* between the faces of a crystal. Huygens had seen that this implied that the crystal must be built of identical molecules piled together like shot, or, as Newton called it, 'in rank and file'. It was, however, left to a retiring French abbé, Haüy, in 1800, to generalize these observations and to show the ways in which these molecules could be associated in different kinds of crystal. It was later found by Mitscherlich (1794–1863) that similar compounds had nearly identical crystal forms, so that the new science of crystallography could become a useful adjunct to chemistry.

ELECTROLYSIS: HUMPHRY DAVY AND FARADAY

Another adjunct was to be found in electricity. The new electric current (pp. 605 f.) was found to decompose not only water but also salts. Davy in 1807 prepared the new metals, sodium, potassium, and calcium, from the previously undecomposed alkalis and earths, thus completing Lavoisier's scheme and dividing all elements into metals and non-metals. It was found that metallic atoms were charged positively and non-metallic negatively. Faraday indeed showed that the rate of transport of atoms in solutions was proportional to their combining weights, and this, of course, leads logically to the concept of a single common atom of electricity – what we now know as the electron. But that final step was to wait for another seventy years, so strong was the prejudice against imputing atomicity to a fluid (pp. 610 f.).

INORGANIC AND MINERAL CHEMISTRY: BERZELIUS

The electrical theory furnished a simple explanation of how salts were formed by the mutual neutralization of positive and negative charges, and this led, particularly in the hands of the great Swedish chemist Berzelius (1779–1848), to the determination of the constitution of most kinds of inorganic compounds and minerals in the first half of the nineteenth century.

The new non-traditional chemical industry which had started in the eighteenth century now grew rapidly under the double impetus of the new knowledge and the vastly increased demands of other industries, especially the dominant textile industry. It was still, however, undertaken in establishments small enough to permit close working contact between scientists and manufacturers, even when these were not the same person. This new industry provided the link between the mineralogical chemist, interested principally in assays of ore, and the druggist with his concern with vegetable and animal products.[5.3]

ORGANIC CHEMISTRY: DUMAS AND VON LIEBIG

Thus for the first time a firm and permanent economic base was provided for chemistry, far larger and better supplied than the pharmacist's shop of the past, and from this base it was possible to build out into the more difficult realms of organic chemistry. Nevertheless, in spite of the genius and ability of the workers in this field, this proved to be a very slow process. Actually the extraction and purification of most simple organic substances such as oils, sugars, and vegetable acids were relatively easily achieved; so was their analysis in terms of the newly known elements, carbon, nitrogen, oxygen, and hydrogen. But the figures obtained by themselves did not tell much – they needed a new kind of interpretation.

This was the work of the new chemists, first in France with Gay Lussac (1778–1850), Laurent (1808–53), Gerhardt (1816–56), and Dumas (1800–84); then in Germany with von Liebig and Wohler (1800–1882). It was Liebig more than any other who restored the primacy of Germany in chemistry after nearly seventy years of French predominance. His laboratory at Giessen was to be the model for the modern chemical teaching and research laboratory. Gradually, from the study of simpler substances – fats, fatty acids, and alcohols – ideas of *structure* began to emerge. As a consequence of a fiasco at a ball where new patent candles bleached with chlorine emitted a frightful stench, Dumas, who was asked to investigate, found that chlorine could be substituted for hydrogen and was led to a general theory of *substitution*. From that followed a theory of *types* of molecules, like alcohols, with some part in common, and then of *radicles*, the detached parts themselves like methyl or benzoyl, which could play the part of atoms.

Such structure could, of course, be merely additive, though already in 1823 von Liebig had found a case of *isomerism* – two substances with the same composition but with different chemical properties. This pointed clearly to some difference of arrangement inside the molecule, but such ideas were firmly resisted, mostly for metaphysical and philosophic reasons. The hypothesis of atoms was not acceptable to a large number of scientists. It seemed to some to go far beyond what experience

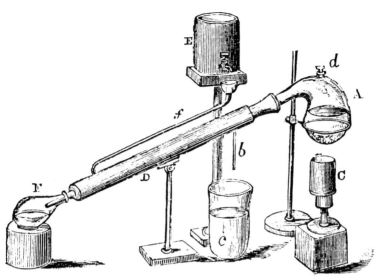

193. The method of condensing vapours obtained during distillation, devised by Justus von Liebig (1803–73), was an efficient practical laboratory method that also had industrial applications. From Henry Roscoe, *Lessons in Elementary Chemistry*, London, 1871.

showed; to others to smack of radical deism. There was also a strong reluctance to believe that substances formed by living beings could be made in the laboratory.

AVOGADRO'S LAW

Organic chemistry might have remained a classified collection of identified substances with mass formulae, and of reactions for turning some kinds into others, had it not been for two impacts from the physical sciences. The first was the recognition of a law originally put forward as early as 1811 by Avogadro (1776–1856), but not generally recognized until restated by Canizzaro (1826–1910) in 1860. This law states that equal volumes of all gases under the same conditions contain the same number of molecules, thus enabling the correct number of each kind of atom in a molecule to be determined.

ASYMMETRICAL MOLECULES: PASTEUR

The second discovery was that of the separation of racemic acid into two components, one ordinary tartaric acid, the other chemically identical

194. Asymmetrical crystals investigated by Pasteur: in 1848 Pasteur showed that racemic acid could be separated into two components, both chemically the same (tartaric acid) but physically different. The physical difference was observable in the way the components behaved in a beam of polarized light. The crystals, too, were different, as illustrated here: the positions of the minute hemihedral faces x, y and z were not the same. This asymmetry is due to the asymmetric nature of the molecules of the tartaric acid and was to prove a discovery of great importance in nineteenth-century chemistry. From J. D. Bernal, *Science and Industry in the Nineteenth Century*, London, 1953.

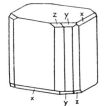 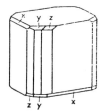

but physically different. This discovery, which was to prove of central importance for the science of the nineteenth century, was made in 1848 by Pasteur, then a young man of twenty-five.[5.3] He showed that whereas molecules made by ordinary laboratory processes did not rotate the plane of polarized light, those naturally produced did so. The former consist of two kinds of molecules of opposite configurations, like right and left hands, in equal numbers; the latter contain only one kind of molecule.

From this critical observation two very different consequences followed. The first was that molecules possessed a shape in three-dimensional space, in other words it was possible to picture them as solid models. The second was that Nature must set about making molecules in a different way from the chemists of that time, and further that there must exist in organisms definite chemical structures which were right-handed, let us say, and not left-handed. Pasteur himself followed the line given by the second clue, which set him among the founders of modern biochemistry and later of bacteriology.

KEKULÉ AND THE BENZENE RING: VALENCY

It was along the other branch that lay the future of organic chemistry, although it was still very slow to evolve. The idea that molecules could be pictured as patterns of atoms in space had been evolved logically by the brilliant German chemist Kekulé (1829–96), who in 1865 conceived the idea as he sat on top of a London bus, that the benzene molecule C_6H_6 contained a ring of six carbon atoms:

From then on it was no longer sufficient to give the numbers of atoms in the molecule of a substance, a mere accountant's description; but to indicate by some kind of plan – an architect's conception – how they were arranged by means of a *structural* formula. Thus he added a decisive proof to the idea that was gradually dawning, that different kinds of atoms were characterized by the number of links that they could make with other atoms. Hydrogen had one, oxygen had two, nitrogen had three, carbon had four of these links or *valencies*.

VAN 'T HOFF AND LE BEL: SPATIAL CHEMISTRY

It was not until twelve years later that simultaneously van't Hoff (1852–1911) and Le Bel (1847–1930) realized that the four carbon valencies could not lie in a plane but must stick out in space, and thus made it possible to explain the two different, right- and left-handed, configurations that Pasteur had discovered twenty-five years earlier. From now on three-dimensional structural organic chemistry became a branch of applied geometry and it was possible both to analyse and to synthesize very complicated compounds.

SYNTHETIC DYES AND THE GERMAN CHEMICAL INDUSTRY

Even before this, however, organic chemistry had established itself in a practical way. Almost by accident Perkin (1838–1907), seeking to make a substitute for quinine, had discovered in 1856 the first artificial aniline dye, magenta, finding at the same time an outlet for the products of coal tar from the gas industry. Chemistry in England, however, was still the pursuit of a few amateurs and even fewer academic university departments, while the chemical industry was proud of being 'practical'. Perkin's discovery, neglected in Britain, was taken up immediately by the more scientifically-minded directors of the new German industry, and the rapid profits accruing from synthetic dyes were ploughed in to create an enormous and dominating German chemical industry. This, though at first ancillary to the textile industry, was, through its capacity for the production of nitric acid for use in the new explosives, to provide the sinews for both the First and the Second World Wars.

The chemist, particularly the chemist of the latter part of the nineteenth century, was effectively a new kind of scientist, one much more

195. A photograph of members of the Chemical Section at the British Association, Oxford, in 1894. Among those present (long row, standing) W. H. Perkin (second from left), Ludwig Mond (fourth from left), William Ramsay (fifth from right). This was the meeting at which Ramsay announced his discovery of the inert gases.

closely tied up with industry than the physical scientist of earlier times. The tendency to identify science with industrial interests which this brought about was one of the major factors that led to the general toning down of scientific controversy, especially of radical scientific attitudes, at the end of the nineteenth century.

From the purely scientific point of view, however, the establishment of molecular constitution by the methods of organic chemistry is one of the greatest achievements of the human mind. The decisive steps were made by very few men, but they were followed by a great crowd of chemists who, using the logic of chemical transformation, were able to imagine the most complicated patterns of atoms in space and actually to make substances having those patterns, thus proving by *synthesis* what they had previously established by *analysis*. In this way organic chemistry grew up as a discipline almost independent of physics, having its own rules and its own way of working.

PHYSICAL CHEMISTRY

This, however, did not apply to the whole of chemistry, especially on the inorganic side, where interest began to shift from the actual composition of bodies to their modes of reaction with each other, to the influence of heat, to such questions as solution, crystallization, and electrolysis. From these interests there grew up a new branch of chemistry ultimately to become a new subject, *physical chemistry*. This was the first hybrid science, one which was to be the prototype of other 'bridge' sciences that in the twentieth century were to link all science into one effective unity. The value of physical chemistry began to make itself felt when attempts were made to exploit industrially the new deposits of mineral salts, particularly the great salt deposits of Stassfurt, which could not be disentangled into their components economically without these methods. It was also the basis of whole new chemical industries such as the Solvay ammonia soda process, which replaced the Leblanc process for the manufacture of soda, and the catalytic processes on which the manufacture of sulphuric acid and of ammonia were based. These were the processes which were to be the main basis of the greatest chemical monopoly concern in Britain.

EARLY BIOCHEMISTRY

The new organic chemistry had another essential part to play in the history of science – it was to lead to a fuller understanding of biological processes. In fact, the beginning of any deeper understanding than the microscope could provide was totally impossible without a knowledge of the laws of combination and the types of structure actually to be met with in biological systems. The nineteenth-century development of organic chemistry had to precede logically any attempt to formulate a fundamental biology.

The main features of animal and plant metabolism as far as carbon, hydrogen, and oxygen were concerned – that is, as far as an animal can be treated as a heat-engine – had been established in the eighteenth century; but it took much of the nineteenth century to establish the equally important role of nitrogen. It was the work of von Liebig that showed what kind of food – nitrogen, phosphates, and salts – plants drew from the ground. The great *cycles* of transformation of the elements, such as that of nitrogen from plants through animals back into the soil, were traced out and even followed into the air with nitrogen-fixing organisms. This was still a far cry from understanding the functions of these inorganic substances in the organism. It is one thing to study the properties, mainly the industrially useful properties, of materials derived from once-living sources; it is quite another to follow them in their transformations during metabolism. That is why organic chemistry took so long to be transformed into *biochemistry*. Nevertheless, as the century came to a close, chemical interest began to shift from the immediately profitable synthetic chemistry of the dyestuffs industry towards an understanding of the more detailed structure of organic substances of a natural kind. This is shown particularly in the great work of Emil Fischer (1852–1919) on the sugars and on the matter of life – the proteins – which he was able to show consisted of chains of much simpler compounds, amino acids. There also, as a by-product of dyestuff chemistry, was laid the beginning of a new chemical pharmacology in the provision of remedies, such as Ehrlich's (1854–1915) salvarsan for syphilis and Bayer 206 for sleeping sickness, which were to foreshadow the triumphs of chemotherapy of the next century (pp. 879 ff.).

9.5 Biology

With the development of physical science throughout the eighteenth and nineteenth centuries, and interacting with it at many points as we shall show, came a renewed approach to a scientific understanding of living things. The roots of this lie much farther back in classical times with the natural history of Aristotle and the physiology of Galen. After a long interval of purely formal and moral interest in Nature, as symbolized in the bestiaries and herbals, interest revived in the increasing pictorial naturalism of the late Middle Ages and the Renaissance, spiced as it was by the wonder and anticipated riches of the New World. Anatomy and physiology were, as we have seen, revolutionized in the sixteenth and seventeenth centuries, and another new world of the very small was opened up by the first microscopists (pp. 467 f.).

The lively interest, however, of the seventeenth-century pioneers in biology, as in physics, turned, towards its end, on one side to a dilettante amusement in the curiosities of natural history, and on the other to the service of a pedantic medicine which included a study of botany and zoology primarily as sources of drugs. This period of discursive observation was, however, to be a very necessary stage in the history of biology – a science incomparably richer in detail than physics or even chemistry, and one consequently where almost innumerable facts had to be collected, examined, and arranged in order before any sense could be made out of them, a task which was to take over 200 years.

The main drives that determined the direction of biological interest, and with it of biological progress, in the eighteenth and nineteenth centuries, were first, those of geographical exploration, largely undertaken in the hope of finding and exploiting new natural products; secondly, the needs of an awakening medicine, with its emphasis on physiology and anatomy; thirdly, the needs and problems of the agricultural revolution which accompanied the transition from traditional subsistence farming to commercial farming for a market; and, lastly, the needs of vastly expanded industries, including those of textiles, food, and drink, largely dependent on animal and vegetable products and, by the very scale of their operation, unable any longer to rely on tradition. These interests all overlapped and interacted. The first two drives remained all the time, though exploration fell and medicine rose in relative importance. Scientific agriculture did not come in till the late eighteenth century, and industrial biology not till the middle of the nineteenth.

Compared with the interests in physical and chemical science, where, as we have seen, there were a limited number of problems set by the advance of industry itself, biology was pursued in a widely scattered and almost casual way. Less able to establish its advances by processes of practical utility, it was necessarily more easily influenced by currents of thought outside science, and particularly, over the whole of the period, by the great religious and anti-religious battles that in different forms convulsed both the eighteenth and nineteenth centuries across the great divide of the French Revolution.

The religious hoped to regain in the animate world the justification for divine governance that had been lost among the celestial spheres. The rationalists hoped, on the contrary, to expel the spirits from the universe by demonstrating the mechanical operation of matter in the phenomena of life, and to explode once and for all the naïve myths of Old Testament creation. Naturalists of both convictions diligently searched Nature to pile up more convincing proofs of what they were certain must be the only right view. Religious preconceptions had no longer the power to prevent research, but they did, at least until the triumph of Darwinism, hold up its most obvious implications. Every inch of the way to the rational interpretation of the world of life had to be fought for, and the only consolation is that, perhaps for that very reason, if it took longer to establish, it was the better understood.

It is in biology, more than in physics, though less than in the social sciences, that men have embraced, often at the same time, the rival stupidities of the commonplace and the marvellous. On the one side everything in Nature is obviously natural – no explanation is needed as to why the grass grows or the lions roar. It is their nature to do so; they always have, and they always will. If from the evidence of fossils or from the tradition of a creation it is admitted on the contrary that the world as we know it must have once been different, it is far easier to believe that it began with a bang, or at most in seven days and from nothing at all, than to attempt to trace its rise step by step from something unfamiliar but not radically different from what we see today. Right up to 1859 the most practical and common-sense naturalist or geologist was quite ready to admit• universal catastrophes, compared to which Noah's flood would be a minor incident, without any mental uneasiness.

In any case, from its very complexity, the great generalizations of biology could only be established on the base of a most extensive and intensive exploration of living things, and this was in the first place to be the task of the natural historians. In what follows we will first trace the development of natural history and its companion study, geology, to its culmination in the theory of organic evolution. For all its importance in the history of human ideas, this great theory was based merely on the external appearance and gross anatomy of living and fossil organisms and had few practical consequences. The other approach through the study of the internal constitution of organisms, large and small, begun by the use of the microscope and continued by the methods of chemistry, was much more searching. It was, towards the end of the nineteenth century, beginning to show its promise of practical utility in the curing of disease and the nourishing of crops.

NATURAL HISTORY AND CLASSIFICATION: LINNAEUS

The eighteenth was the great century of travellers, collectors, and classifiers. The idea of classification arose from the practical necessity of arranging plants in botanical gardens, collections in cabinets, and even more perhaps from the making and printing of catalogues. Very naturally, each collector and cataloguer had his own ideas as to how to arrange his material, and the result was a welter of confusion of names and arrangements.

It was only in the middle of the eighteenth century that an energetic and systematic young Swede, Carl Linnaeus (1707–78), afterwards ennobled to von Linné, the son of a poor parson and almost self-taught, took upon himself, at first single-handed, the task of classifying all animals, minerals, and particularly vegetables in the world. In botany, where his chief contribution lay, he had the genius to see that the great discovery of Camerarius (1665–1721), that flowers were the sexual organs of plants, was the key to their classification. Basing himself on the numbers of the hitherto neglected stamens and pistils, Linnaeus divided the plants into classes and orders. For the finer divisions of

196. A portrait of Linnaeus (Carl von Linné, 1707–78), in Lapland dress. From a contemporary engraving.

genera and *species* he established the double-name nomenclature, *Linnaea borealis* L,* which would provide enough actual words to enable every living thing to be distinguished.[5.53]

The time was ripe for such an organization of knowledge, even if quite arbitrary – for at first it was little more. Linnaeus travelled widely, collected copiously, and built up a systematic botanic garden at Uppsala. He soon attracted a band of devoted disciples who travelled all over the world to complete his classification, and found everywhere admirers and imitators. The Linnean Society of London was founded in 1788. On the basis of his simple system and his undoubted mastery of the material, Linnaeus imposed his classification on the whole of the learned world. With later modifications it remains that of botany and zoology to this day. On the other hand, his classification of minerals, being based, inevitably for the time, on unscientific principles, was soon abandoned and gave way to the more rational system based on chemistry and crystallography.

TOWARDS A NATURAL SYSTEM : BUFFON

Armed with this system, the naturalists, in whatever part of the world they were, could work together knowing that if they got the name right they were talking about the same organism, and that thus they could contribute to building up a common catalogue of organized beings, a process which is continuing to our day. The Linnean system was too rigid to start with; but it was possible, without substantially breaking it up, to alter it progressively till it became more and more a natural system: till species that resembled each other more than any others appeared together in the same genus, while the larger groups, genera and families, were divided from each other by more important differences.

The work of the systematists had immediate and lasting practical value. Scientifically, however, its establishment had a much more far-reaching consequence. It was impossible henceforth to contemplate the natural classification of living things without being forcibly reminded of their relationships, which were indeed implied by the very terms used, those of genera, or tribes and families. One of the first to sense this was George Louis de Buffon (1707–88). By his brilliance and affability he did more than any other man to popularize natural history both at the French Court and among the rising bourgeoisie, of which he was an ennobled member. In 1739 he was made keeper of the Jardin du Roi, now Jardin des Plantes, and turned it into what was, for the time, a great research institute, where many of the biologists and chemists of France received their inspiration and training (p. 534). Unlike Linnaeus, who

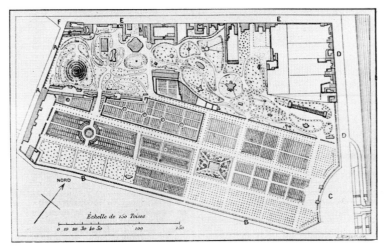

197. Ground plan of the Jardin des Plantes (Jardin du Roi) from an 1821 engraving.

lacked any other knowledge than that of natural history, Buffon was originally a physicist and brought the rational ideas of the Newtonian synthesis into the field of biology; he was, however, possibly for the same reason, by no means as patient an observer or as diligent a classifier. He set the fashion in the literary presentation of science, and his daring ideas on the origin of the world, of plants, animals, and of man himself, endeared him to the *philosophes* and the makers of the French Revolution.

Buffon, in his monumental *Système de la Nature*, claimed that the relationship implied by the classification of animals and plants was a real one. In this he was supported by Erasmus Darwin (1731–1802), who has already been mentioned as a leading member of the Lunar Society. He was a successful doctor of Lichfield, a poet, a popularizer of science, and a speculative and daring theorizer in biology. His *Zoonomia* was an attempt to trace the origin of life from a primitive filament, which produced the great variety of observed living forms as a consequence of its different reactions to a variety of external influences. As he could not have had any knowledge of either the intimate structure of living matter or of the mechanism of its reactions, his ideas were necessarily speculative, and served more immediately to support the *Naturphilosoph*–romantic school of Germany (pp. 645 f.) than to lead to any new observations and experiments. Still, what he dared to think, others, with better grounds, could think after him.

If it had not been for the pietistic reaction against the French Revolution, the idea that all species came from a common stock would have been freely accepted early in the nineteenth century. However, almost more than in the seventeenth century or in the time of the Counter-Reformation, it was necessary in the early nineteenth century to uphold the literal truths of the Bible stories of the creation of the species, of animals and plants, on the appropriate days, so most naturalists for over fifty years put their blind eye to their microscopes and refused to think about the meaning of the system of Nature.

EARLY EVOLUTIONISTS : LAMARCK

In spite of this, some continued to speculate, of whom the most original, Lamarck (1744–1829), botanist at the Jardin du Roi, boldly propounded the theory in 1809 that the species of today were derived from those of previous times by an adaptation brought about by their desire to fit more closely with their environment. The giraffe, seeing leaves growing on a high tree, stretched his neck, and that stretching was inherited by his descendants. The idea seemed far-fetched and won little support, but meanwhile the evidence was accumulating, and not only from the study of living organisms, but now even more from that of fossils.

SPECULATIVE GEOLOGY AND CREATION

The study of geology came late into the category of sciences. It was pre-eminently a field science. The collector in his cabinet could do little but marvel at the odd productions of the earth. The miner, on the other hand, was so concerned with the ore and the indications of its presence in other rocks that he had usually neither the inclination, nor the learning, to formulate any general theories as to the structure and history of the earth. Yet speculation about the earth and its fossils grew steadily in the eighteenth century with the general increase of interest in Nature. In fact, from even earlier times, the idea that the shells found in mountains implied the presence of sea had led to speculations about the antiquity of life, though in the past the whole matter had been easily smoothed over by putting it all down to Noah's flood. From voyages and astonishing accounts of foreign volcanoes and earthquakes another view began to be held: that the world was subject to continual cataclysms in which the crust was broken up by internal fire; and the controversies between the neptunists, or flood-believers, and the plutonists, or earthquake-believers, remained an unprofitable exercise throughout the latter part of the eighteenth century.

HUTTON AND COMMON SENSE

The first radical break from speculative geology came from Hutton, an Edinburgh doctor, a close crony of Black's, and one of the great company of brilliant scientists and philosophers who had made their city the Athens of the North (p. 530). In his *Theory of the Earth* (1795) he put forward the idea, revolutionary in its common sense, that the phenomena of geology are the products of forces that we still see acting round us. From his walks in the country, from his experience as a practical farmer, he concluded that valleys were cut by rivers, and plains deposited out of the mud they brought down, which then hardened into rock. He also understood that the massive unstratified rocks of Arthur's Seat could not have been deposited from water, as the arch-neptunist, Werner (1749–1817), maintained, but must have been formed from the solidified lava of an ancient volcano. These views were too rational to survive the reaction against the French Revolution, which brought in a school of geologists, often in holy orders, who were everywhere looking for vestiges of creation,[5.52] but Huttonian views never entirely disappeared.

The success of field geology came from the experience of cutting canals rather than from the intensive and highly localized craft knowledge of the miners. William Smith (1769–1839), the practical surveyor and canal-maker, realized from his work that throughout the whole of southern England the layers or strata of the earth lay over one another in one invariable series, and he spent most of his life plotting their outcrop in the first of geological maps.

LYELL'S 'PRINCIPLES'

Catastrophic theories of how these strata came to be laid down became more and more difficult to maintain. They were quietly dropped when Lyell in his *Principles of Geology*[5.71] revived Hutton's doctrines of the operations of natural forces, and founded his *uniformitarian theory*, based on far more extensive observations.[5.18] But if every stratum represented a deposit of a certain age, the distinctive fossils it contained must have belonged to animals living at that time, and these fossils corresponded to quite different forms of life, and even showed definite progressions. Reptiles, for instance, did not appear before the secondary strata, or mammals before the tertiary. Lyell, accepting as a logical necessity the fixity of species, could only deduce that a whole new fauna had been created at every geological age and had become extinct in its turn. Obviously all this must have taken a very, very long time, so that the Bible

198. In his *Principles of Geology*, published in London in a number of editions from 1830 onwards, Charles Lyell (1797–1875) advocated the idea of slow non-catastrophic geological change to account for the Earth's surface features. This frontispiece from the ninth edition, 1853, shows the Temple of Serapis in 1836, illustrating its slow subsidence.

story of the Creation, quite apart from its miracles, became increasingly difficult to believe. Yet, in the atmosphere of reaction in the early nineteenth century, it was extremely daring to question it.[5.52]

CHARLES DARWIN AND ORGANIC EVOLUTION

In fact, it was not until the evidence was altogether overwhelming, and there was in addition some plausible mechanism to explain how different kinds of animals could have descended from each other, that it was possible to break the spell of ancient religion. To provide for that mechanism in the form of natural selection was to be the work of Charles Darwin, the grandson of Erasmus Darwin. He was a typical by-product of mid-Victorian capitalism, a man of independent means who, after a formative voyage round the world in the *Beagle*,[5.36] could settle in his study and garden at Down House and survey minutely and carefully aspects of animate Nature bearing on the problem of the origin of species.

Darwin had been particularly impressed by the species problem in his study of the distribution of rare species on isolated islands such as the Galapagos. It was very tempting to imagine that such species had come from ancestors on the mainland and had somehow grown different – but

199. 'Beagle Laid Ashore. River Santa Cruz.' The Beagle, commanded by Captain Robert FitzRoy, voyaged round the world between 1831 and 1836. The naturalist on this expedition was Charles Darwin, and the work he did on this voyage laid the foundations for his *Origin of Species* (1859). From FitzRoy, *Narrative of the Surveying Voyages of His Majesty's Ships Adventure and Beagle*, London, 1839.

how and why? Might it have something to do with the conditions of their life which favoured some features more than others? He began to think that possibly the conditions of competition of human economic life might also apply to the animal world. Indeed a fully elaborated theory built to justify capitalist exploitation was ready to his hand. Life, according to parson Malthus (p. 1060), was a struggle in which the best survived, and wealth and position were the rewards of virtue in this struggle. Disease and war were the means by which a population was kept from becoming more numerous than the available food supplies could support. If the same were to happen in animal societies, thought Darwin, those that varied ever so little in the direction of being more fit for their environment would pass on that advantage to their descendants, and so, gradually, the species as we know them now would evolve. The hungry forties was a very suitable time to observe this phenomenon taking place.

NATURAL SELECTION

Darwin, however, was a most cautious man and did not publish this idea. Instead he spent nearly twenty years building up the evidence for it. He drew it from all sides of natural history – from the record of the rocks, which showed the gradual elaboration of form in previous ages; from the distribution of animals and plants in the world; and finally from the study of the great breeding experiments that were going on in the nineteenth century, partly to improve stock and partly for fancy breeds of dogs or pigeons, which provided him with examples of changes as strange as any that occurred in evolution. Yet he might not have published his theory at all even then if another and much younger traveller, Alfred Russel Wallace (1823–1913), had not independently come to the idea of evolution of species through his study of the distribution of animals in the East Indies.

THE 'ORIGIN OF SPECIES' AND THE EVOLUTION CONTROVERSY

The explosion that followed the publication of the *Origin of Species* showed how prudent the retiring Darwin had been in holding back his ideas. Even in the relatively advanced sixties they were to create a prolonged and bitter controversy. This, however, was to turn on questions of a theological or political rather than of a purely scientific nature (p. 663). In biological science it produced an enormous effect of liberation. It provided a unifying principle for the whole living world.

However, the effect of Darwinism on science was not an altogether happy one. It certainly did raise a great interest in biology and drew many people into it. But at the same time the emphasis that Darwin's

200. Darwin's finches: in his voyage to the Galapagos Islands, Darwin was astonished to notice variations in a single species, although environmental conditions were similar and had led him to expect identical creatures on every island. In his *Journal of Researches*, 1845, he showed the differences observed in the beaks of ground finches and correlated these with their feeding habits. They are (1) *Geospiza magnirostris* which feeds on large seeds; (2) *G. fortis* which consumes smaller seeds; (3) *Camarhynchus parvulus* which feeds on insects; (4) *Certhidea olivacea* which eats small insects. From *Charles Darwin* by Gavin de Beer, London, 1963.

theory gave to the simple tracing of evolutionary relationships between organisms and the building of elaborate family trees distracted naturalists from the study of the actual lives and of the inner workings of animals and plants. For this no one could blame Darwin himself, who was, as his detailed researches on such varied topics as earthworms, carnivorous plants, and the expression of the emotions show, one of the pioneers of experimental biology.[5.33-5]

NATURPHILOSOPHIE

The sequence of the species controversy has been followed down to the end of the nineteenth century. It is now necessary to go back to the beginning of the century to pick up another thread to the understanding of living things from the study of their structure. Here again much of the initial impulse came from natural history, but more from the aspect of anatomy and physiology with its close relation to medicine.

It was especially in biology that the mystical trend in science – Neoplatonic, Lullian, Paracelsan – found its last serious expression in the German *Naturphilosophie* of the early nineteenth century. Inspired by philosophers like Herder and Schelling and by poets like Goethe, there came the search for the Absolute Idea or the Divine Plan of Nature, which was also incomprehensibly bound up with the regeneration of the German people and the destruction of abominable French mathematical materialism.[5.82] Nevertheless, the search for the Archetypes meant a comparative study of the structure or *morphology* (the word is Goethe's) of animals and plants, which was to continue long after the ideas that had given rise to it had evaporated. Lorenz Oken, already mentioned as a refounder of German science, was one of the finest representatives of this school, and was responsible for the recognition of the common features of the structure of the main groups – phyla – of organisms living and extinct (p. 550).

THE MICROSCOPE:
TISSUES AND CELLS

Besides this naturalistic approach there was another from medicine. Though the amalgam of classical Galenical medicine with its Arab commentaries still dominated medical practice, the old theories of medicine based on the humours could not stand in the face of the advance of chemical and biological science. Yet even as late as the beginning of the nineteenth century there could be nothing effective to put in their place. The result was an era of wild speculation and system-building in which inspired quacks, like Mesmer with his animal magnetism, and over-confident anatomists, like Gall with his phrenology, gained a wide following.

At the same time, the renewed interest in anatomy and physiology was to lead to the greatest advances since the Renaissance in the understanding of the body in health and disease. Bichât (1771–1802), in his short life, virtually refounded *pathology*, and by the careful study of the structure of the different organs distinguished the *tissues* – nervous, arterial, venous, muscular, fibrous, glandular, epidermal – that were common to many of them. This study was followed by others in which the new achromatic microscopes of Amici (1827) enabled a far greater insight to be obtained into the fine structure of tissues, *histology*, than had been possible to the pioneers of the seventeenth century. This revealed that the tissues in turn were composed of cells – square cells for the liver, long cells for the muscles, enormously elongated cells for the nerves.

THE CELL THEORY

The whole body, as Schleiden (1804–81) and Schwann (1810–82) pointed out in 1839, could be treated as a colony of *cells*, and, what is more, all had arisen from one, or rather, two cells; the cell of the egg and the cell of the sperm. The actual growth of the organism from the fertilized egg-cell had been followed out by von Baer (1792–1876) at about the same time. The new science of *embryology*, which he virtually founded, also brought out the kinship of different animals in each great group or phylum such as the vertebrates. The cell theory made intelligible the growth of the individual, just as natural selection was to make intelligible the development of the species; and both seemed to be following a parallel track of evolution. The use of the microscope in all fields of biology began to reveal unsuspected complexities, but in the early stages they had very little effect on actual practice. It was only when the simplest kind of animals and plants, the fungi [5.69] and the simpler single-celled protozoa and bacteria, came to be studied that some understanding of the life and functions of cells was reached, and with it the possibility of the control of living organisms.

FERMENTATION

As often happens in the history of science, this achievement was to come from right outside biology, from the study of agricultural pests and of industrial chemistry. From before the dawn of civilization men had made use of processes generally known as fermentation when the result was pleasant, or as putrefaction when it was not. By careful practice and exact following of rules they had even managed to secure a definite and reproducible control of certain limited sets of processes – the brewing of beer, the maturing of cheese, or the tanning of skins; but, like all technically achieved processes, it was extremely difficult and very dangerous to change them, and the enormous expansion of demand that was created by the new populations of the early nineteenth century made not only for expansion in consumption but also for numerous disasters.

PASTEUR AND BACTERIOLOGY

It was in the growing industrial town of Lille in 1855 that the young professor of chemistry, Pasteur, first came into contact with the activities of living ferments. The beer and the vinegar, usually good, would sometimes unaccountably go bad, and Pasteur, finding no chemical explanation, looked at them through the microscope. He found that when normal fermentation went on there were the little round cells of yeast, already studied in 1839 by Caignard de la Tour (1777–1859), but abnormal fermentations were characterized by different organisms, what he called the vibrios, because they kept on dancing continually in his field of view.[5.91]

201. Louis Pasteur (1822–95) in his laboratory at the Ecole Normale, Paris. From the *Illustrated London News*, 1884.

Now Pasteur had already, as we have seen (pp. 629 f.), been concerned with the chemical activity of living things in producing asymmetric molecules. His experiments with moulds had convinced him that the processes of fermentation themselves must be due to living organisms and not to any inert chemical reactions. As a chemist he studied not only the appearance of the micro-organisms but also their chemical performance. He investigated whether they could live in or out of the air, and as a result was able to devise ingenious but practical ways, including the process now known as pasteurization, of preventing them interfering with the successful production of beer or vinegar.

It was his knowledge of the living organisms in fermentation that spurred Pasteur on in his vigorous denial of the possibility of spontaneous generation of life and which led to his famous controversy with Pouchet (1800–72). There he showed that by excluding the invisible *microbes* of the air, animal and vegetable substances could be kept indefinitely without putrefying. He thus convinced the learned world of the facts which the chef, Appert,[5.14] as far back as 1810, had made use of in his method of preserving foods by boiling them and sealing them in glass vessels, which was later to be the basis of the great canning industry. It had, however, been objected that Appert's bottles contained no oxygen, which was claimed as the cause of putrefaction. Pasteur had to show that the filtering of air was equally effective in preventing putrefaction.

Pasteur's preoccupation with the organic side of fermentation also brought him into opposition to von Liebig's view that it was due to a specific chemical ferment, and his success pushed this into the background. It was only in 1897 that E. Buchner (1860–1917), almost by accident, isolated such a ferment from ground yeast and inaugurated the study of *enzymes*. Thus in the end both von Liebig and Pasteur were proved right. Fermentation is brought about by a ferment, but that ferment can only be elaborated by a living organism (p. 887).[5.3]

THE SILKWORM DISEASE AND THE GERM THEORY

In 1865 Pasteur was called to a more difficult task. The new industries of France depended very largely on the supply of silk, and this was threatened with extinction by a mysterious disease of the silkworms. Pasteur was sent to deal with it. At the time he was so little a naturalist that he did not even know what a silkworm was or that an ugly caterpillar later turned into a beautiful moth. Nevertheless, after a season's intense research he found that the disease was due to a kind of organism that actually lived and grew inside the caterpillar itself. This provided the clue to wiping out the disease.

From then on he came to think more and more that the diseases of larger organisms, of animals and of men, were due to similar causes, to the minute germs of disease. This was not a new idea. In effect, it was as old as disease itself, and the phenomena of infection and epidemics bear witness to it. Jenner (1749–1823), indeed, had long before taken the first official practical step to control smallpox by *vaccination*, which presupposed the presence of an active *virus* of disease in a milder form in contrast to the drastic *inoculation* with smallpox itself which had been practised for centuries. But these *germs* of disease could never be recognized, and the medical profession, into whose Aristotelian or even Hippocratic theories they did not at all fit, refused to admit their existence. Yet they had been seen years before by Leeuwenhoek by means of his simple but excellent microscopes. But there seemed to him no obvious connexion between the minute creatures that he saw and the diseases that afflicted animals and men (pp. 467 f.).

When the evidence had accumulated on both sides for 200 years the discovery of the role of bacteria became overdue. As in similar cases, Pasteur was neither the first nor the only one to make it. Koch (1843–1910), a German country doctor, following Davaine (1812–82), studied the multiplication of the anthrax bacillus, and developed the method of growth on gelatine which made it possible to obtain pure strains – a method he used later to isolate the agents of tuberculosis and cholera.

202. In 1865, Pasteur's studies of the diseases of silkworms led him to the concept of the disease-producing activities of micro-organisms in animals and man. From Pasteur's *Études sur la maladie des vers à soie*, Paris.

VERS SAINS.

a Ver sans tache. b Ver avec taches de blessures

Lister (1827-1912), in Scotland, developed the practical techniques of antisepsis that began to cut down the appalling mortality in hospitals. Pasteur was, however, the main standard bearer in the war against the microbes.

PASTEUR AGAINST THE DOCTORS

More by his devotion to the good of mankind and his terrific force of character than by cold scientific argument, Pasteur succeeded in breaking down the opposition to this new approach to disease, for it was a very furious opposition, comprising almost the whole of the medical profession. He needed all his early reputation as a chemist, all his acquired reputation as an industrial adviser and as a conqueror of the silkworm disease, before he could persuade the authorities of the various hospitals to adopt what are now considered the most elementary precautions of asepsis. But once he had demonstrated his results of immunization, first for anthrax in cattle, and lastly and most spectacularly for rabies in man, popular enthusiasm forced even the doctors to accept his ideas.

THE FOUNDATION OF SCIENTIFIC MEDICINE

The revolution introduced by Pasteur was effectively the foundation of scientific medicine. In previous centuries much had been found out about the body and its behaviour in health and disease, but this was only a half-science, capable of prediction and palliation of symptoms, but lacking the telling proof of controlling disease by effective prevention or cure. The few methods of prevention such as quarantine and vaccination, or of cure like mercury for syphilis or quinine for malaria, had been intelligent utilizations of chance discoveries or tribal traditions. But because they were not based on any scientific theory they could not be generalized and used to cure other diseases. Without the germ theory it was impossible to understand what was happening in acute infectious diseases, and doctors had to let them run their course and even helped unwittingly to spread them.

THE CONTROL OF EPIDEMICS: BACTERIOLOGY

Once the germ theory and the technique were clearly grasped, dozens of devoted men could study an infectious disease in the field, track down the causal germ, and often, though not always, find an immunizing or curative serum, and even without this could indicate the precautions necessary to stop epidemics. Checked by improved sanitation, water-borne diseases such as typhoid began to disappear from Europe and the child-killing diphtheria to diminish. In turn the great scourges of cholera, plague, and malaria were controlled, except where poverty made the new measures impossible to apply.

The very success of the germ theory of disease in showing the way to control most of the acute diseases which decimated mankind in childhood and youth blinded public opinion, and to a lesser extent even the profession of medicine for a time, to the fact that only the advance guard of disease had been driven back, and that in treating disease, as externally caused, the reactions of the body were being neglected. There still remained the crippling disease of rickets, and the killing diabetes, heart disease, and cancer, to challenge the scientists of the next century. Nevertheless, through *bacteriology*, science had once and for all entered the field of medical practice and was soon to become an integral part of medical tradition.

The work of Pasteur and his pupils, and of the other schools of bacteriology, meant much more for science than its immediate medical results, critical as they were in the history of civilization. He had, by his earlier work, already demonstrated that even the simplest of creatures did not arise *de novo*, that no creation of life on this earth was still going on. That these tiny organisms were alive seemed certain by their movements and reproduction. But their life must be a very different one from that of the higher organisms, a life that was essentially chemical rather than mechanical – dependent on molecular rather than on bony architecture. He was thus one of the great forerunners of the biochemical revolution of the twentieth century.

CLAUDE BERNARD AND PHYSIOLOGICAL CHEMISTRY

Another forerunner was also a Frenchman, Claude Bernard (1813–78), who studied the physiology of living men and animals and discovered that the important internal activities of the body were carried out by a complex balance of chemical reactions, many of which he unravelled, a balance the maintenance of which was a necessary condition for life itself. The higher the organism, the more it tended to keep its internal conditions constant and independent of the external conditions, and was thus capable of reacting when simpler organisms were frozen into immobility or cooked to death.*

NEUROLOGY

The study of the mechanism of nervous control, an aspect of physiology that had lain dormant since the experiments of Galen nearly 2,000 years before (pp. 222 f.), also came to life again in the nineteenth century. The function of the nerves both in sending messages to the muscles and receiving them from the sense organs was, thanks to the work of Bell (1774–1842) and Magendie (1785–1855), at last understood, and their connexions were tracked out through the vast complexity of the nervous system.[5.109] This threw the first light on the controlling function of the most complex network of all: the brain. Even in the nineteenth century materialist biologists were casting doubts on the absolute nature of pure mental phenomena. Physiology was beginning to reveal how almost infinitely more complex were the bodies of even the simplest animals than anything the philosophers had imagined.

SCIENTIFIC AGRICULTURE

Of the four sources of biological knowledge in the eighteenth and nineteenth centuries already described – natural history, medicine, agriculture, and industry – the contributions of the last two have inevitably been mentioned in dealing with the first and second. Darwin's ideas were much affected by the practical successes of the animal-breeder and horticulturist. The early bacteriologists first secured their successes in dealing with the diseases of animals, and Pasteur himself was led to bacteriology through the industrial processes of wine-making, brewing, and silk manufacture. Nevertheless, there remains an independent stream of scientific thought which stems from the central problems of agriculture: How do plants grow in the soil, and what constitutes the food of men and animals?

From the beginning of the eighteenth century, wherever the capitalist economy had penetrated, the problems of agriculture were brought into the forefront. Venerable tradition no longer served when it was a question of getting the greatest returns from the land. Individual improving farmers banded together with progressive landlords in societies for the promotion of agriculture,[5.4] and in view of the temper of the times it was natural that science should be involved in the task of laying bare its underlying principles. This, however, proved to be a very difficult one.[5.30]

Not until the mid nineteenth century and after many false starts was it possible to go beyond the direct experience of farming practices themselves. It was a matter of trying out variations of existing methods, noting which gave increased yields, and following up promising clues. Great innovations came from industry rather than science in the form of farm machinery, which revolutionized ploughing, sowing, harvesting, and threshing. The steam-engine, however, gave far less to farming than it did to industry or transport. Complete mechanization had to wait for the smaller, lighter internal-combustion engine of the twentieth century.*

THE NUTRITION OF ANIMALS AND PLANTS

It was on the chemical rather than on the biological or mechanical side that science made its most effective contact with agriculture. The pneumatic revolution in chemistry, beginning with Priestley and culminating with Lavoisier, had shown the animal organism as a kind of heat-engine burning food for fuel, and the plant as reversing the process, using sunlight to rebuild living tissue from waste gases and to restore the oxygen to the atmosphere. In Moleschott's (1822–93) classic phrase, 'Life is woven out of air by light.'

None of this, however, could have any practical bearing until the role of the soil was elucidated. The practical farmers and gardeners knew that the soil fed the plants, yet the scientist from 1790 to 1840 was at a loss to know how precisely it did it. Van Helmont had shown, 200 years before, that a willow tree could grow on water alone. It seemed quite

203. Liebig's work in organic chemistry led to commercial developments, of which Liebig's meat extract was but one of many. From an advertisement *c.* 1904.

reasonable then to assume that the element water had been transmuted into the element earth or wood. But after 1790 this was shown to be alchemical nonsense, and there was nothing to take its place till von Liebig's classical investigations. His report on *Chemistry and its Applications to Agriculture and Physiology* (1846), prepared at the request of the British Association (p. 550), established the division of living tissues, and consequently of foods, into the now classical carbohydrates, fats, and albuminoids (proteins). He showed that the first two were primarily fuels formed in plants from the carbon dioxide of the air, and that only the last of them contained nitrogen and were formed in plants from nitrates drawn up from the soil together with other essential elements, such as phosphorus and potassium, to be returned later to it from the excrements of animals in another great cyclic process of Nature.

ARTIFICIAL MANURES

With the elucidation of the chemical role of the soil came the first explanation of the action of farmyard manure, and with it the possibility of supplementing it from other sources. Sir John Lawes (1814–1900), a gentleman of scientific tastes, turned his estate at Rothamsted into the first agricultural research laboratory, experimented with nitrates, phosphates, and potash from various sources as substitutes for farmyard manure, and even built factories to produce them. From this and analogous experiments in other countries came the great fertilizer industry, which in the latter part of the nineteenth century served the double purpose of intensifying agricultural production and supplementing the needs of textile chemicals in building up a highly monopolistic heavy chemical industry ready to supply the war needs of the twentieth century.

THE FOOD INDUSTRY : REFRIGERATION

Parson Malthus considered that:

In the wildness of speculation it has been suggested (of course more in jest than in earnest), that Europe ought to grow its corn in America, and devote itself solely to manufactures and commerce, as the best sort of division of the labour of the globe.[5.76]

Before his jest could be played in earnest it was necessary to send out the men to grow the food in distant lands, whether as slaves, convicts, or hunger-driven emigrants, and to find the means of getting the food back in an edible state. Traditional means of doing this certainly existed – drying, salting, boiling, and freezing go back to the Stone Age – but they could never have been used on a scale adequate to feed tens of millions, if they had not been rationalized and transformed by the infiltration of science.

On the one hand Pasteur's life-work had shown the need to exclude germs, on the other the new thermodynamics showed the way of using a heat-engine in reverse to produce artificial cold (p. 587). Canning and refrigeration between them ensured that food could be made available wherever money could be found to pay for it. It also ensured the domination of the packing and refrigeration companies over all the open spaces of the world where beef could be moved on the hoof. One end of this process has been romanticized in the cowboy and the gaucho, the other is to be found in the stockyards of Chicago or Cincinnati, where the mechanization of slaughter was to provide the prototype of the assembly line of the mass production of the next century.[5.50]

204. 'Over London by Rail' by Gustave Doré, showing the cramped houses and small backyards of the nineteenth-century industrial city. From Gustave Doré and Blanchard Jerrold, *London*, London, 1872.

APPLIED BIOLOGY : MEDICINE AND AGRICULTURE

By the end of the nineteenth century biology had taken its place with the older sciences of physics and chemistry as a rational scientific discipline, though it still retained many of the vestiges of earlier magical and mythical beliefs. Nor had it as yet anything like the understanding and control of its material that the older sciences had already achieved. But it was already proving its practical utility. Indeed, the great economic advances of the latter part of the nineteenth century would have been quite impossible without the help of applied biology. In fact this is one of the best examples of the Marxist dictum that 'mankind always sets itself only such tasks as it can solve'.[5.80.357]

The vast agglomeration of people in nineteenth-century manufacturing towns could never have been maintained without the sanitary methods which were evolved as a consequence of the gradual appreciation of the germ theory of disease. Nor could these populations have been fed without the application of the new chemical knowledge of the nutrition of plants. The use of nitrogenous and phosphate manures was the major factor in the increased productivity of the land, and in the possibility of extending, much farther than had previously been thought possible, the areas of cultivable soil. Finally, the tropical products such as rubber and oil, so essential to the development of industry, could not have been won in the quantities required unless at least the worst of the tropical diseases had been brought under control.

9.6 Retrospect

SCIENCE IN THE AGE OF CAPITALISM

We have now followed in outline some of the main streams of scientific advance in the eighteenth and nineteenth centuries, and have seen connexions both with the material development of society exemplified in the Industrial Revolution and its consequences, and with the evolution of thought which was needed to bring man into effective relation with his new socially created environment. It was in this period that capitalism came fully into its own, flourished most exuberantly, and began to show the first signs of decline. Science also grew mightily and continuously, apart from minor fluctuations, and its growth must have been even more rapid than that of the economy as a whole, for it occupied a far more important position at the end of the period than at the beginning. In

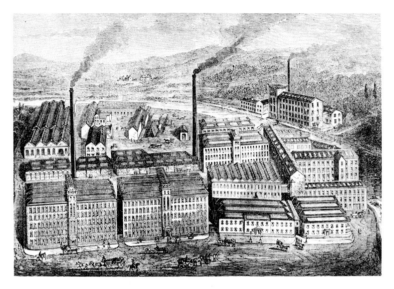

205. By the latter decades of the nineteenth century, the large factory was common-place. J. & J. Clark's Anchor Thread Works at Paisley. From *Great Industries of Great Britain*, London, *c*. 1880.

the early eighteenth century it provided, in the steam-engine, the motive power for an industry that was still largely built on a basis of traditional techniques, and owed much to ingenuity and little to science. Towards the end of the nineteenth century new major industries based entirely on science were arising. In addition, science was permeating the older craft industries and agriculture itself. At the beginning science still had more to learn from industry than it could give to it, at the end the very existence of industry was bound up with science. Through the technical transformation of industry that it had made possible, science was affecting the development of capitalism, enabling it to turn away from the individualist free competition of small-scale industry to the large monopolist undertakings with deliberately planned and scientific production methods.

A comparison between the scientific revolution of the sixteenth and seventeenth centuries discussed in Chapter 7 and the Industrial Revolution of the eighteenth and nineteenth brings out the radical change in the kind of relation between science and economic life. In the first period, as we have seen, the call on science and its effective answer were on a very limited front, hardly more than that of astronomy and navigation. In the second the whole range of industrial activities was included: mechanism, power, transport, chemicals, and munitions. Correspondingly the science of the first period was concerned mainly with new *instruments* for the collection of information about Nature – telescopes, microscopes, thermometers, barometers – and with the mathematical analysis needed to design them and interpret their results. In the second period, though instruments continued to develop and multiply, they were now only a part of the *material* products of science. New machines – steam-engines, turbines, dynamos, electric motors, chemical plant – all designed not just to find out about Nature but to change it, were the characteristic products of the eighteenth and nineteenth centuries.

Between one revolution and the other science had indeed changed from the passive to the active role, from the investigation of Nature to the 'effecting of all things possible'. This transition was made possible, technically by the very development of machinery, largely the fruit of joint efforts of workmen and scientists, and economically by the availability of capital in ever-increasing amounts as the profit from earlier investments accumulated. It is this strictly capitalist mode of financing technical and scientific advance that accounts for the great bursts of activity in the late eighteenth and mid nineteenth centuries.

Compared with any previous era a prodigious effort was expended. It is only when we look at it in terms of the absolute effort of today that it seems so puny.[5.93;5.110] The total amount, for example, spent on scientific research in Britain in the whole nineteenth century cannot have been much more than a million pounds.[5.3] We now spend on civilian research alone some 400 times that amount annually. The links of science with profit also account, as I have shown elsewhere,[5.3] for the highly irregular rate of that advance. Even when an application of science seemed to promise large returns, the lack of available capital for ventures that would lock it up for some years, and could not be guaranteed success in the end, deterred all but the most sanguine entrepreneurs.

THE WORKING CLASS AND SOCIALISM

The capitalists had used science most willingly when it served their purpose for increasing profit. They used it reluctantly and belatedly in applications for the public good, such as health and education. They absolutely refused to use it when it was a matter of examining and possibly altering the system from which they drew their wealth. But if they would not do so, others would. In the process of making science serve profit the capitalists had shown the way to the large-scale social mode of production that would make the profit motive unnecessary. They had at the same time brought into existence a working class to whom the capitalist system stood for toil, insecurity, and want.

At the outset of the period a new emergent capitalism was effectively shaking off the last vestiges of the old feudal system of production and was setting out on a career of progressive expansion. At its end, capitalism, enormously developed, had spread its dominion all over the world, but it now stood on the defensive against a newly risen working class, all but ready to move on to a new and more comprehensive socialist mode of production, and one able to use the results of science to the full (pp. 707, 1179 f.).

In assessing the effects of science on life and thought over the eighteenth and nineteenth centuries, it is accordingly necessary to trace the transition from its liberating effects at the beginning, where it was allied to all the forces of progress, to its ambiguous and uncertain state at the end, where progress could no longer be taken for granted and war and social revolution loomed over the mental horizon. The dividing line came with the French Revolution and the reaction that followed it. For all their patronage, both the old régime in France and the Church and King party in Britain, with their base in landed property, had necessarily to stand against science. The advancement of science accordingly became associated in the later eighteenth century with rising industry, political reform, and liberal theology, serving largely to justify an optimistic and progressive outlook.

After 1815 the position was no longer so simple. Science itself was deeply divided into conforming and liberal sectors, as exemplified for instance in the history of geology and in the evolution controversy (pp. 644 f.). Its old tradition and the practical effects of its discoveries tended to identify science with the great nineteenth-century expansion of capitalism, but the identification was no longer whole-hearted or cheerful. Against it stood the evident fruits of the application of science in the blight and ugliness of industrial areas, and with it the awareness, hostile or conscience-stricken, of the mob, of the new proletariat. The spectre of Communism – however, as yet, ineffective in action – haunted the intellectual as well as the political scene. After 1870 much of the cheerfulness had evaporated and an apocalyptic note crept in.

SCIENCE IN THE WORLD OF IDEAS

The direct effect of science on the ruling ideas of the period was far less important than its indirect effect through its association with the Industrial Revolution; but it was nevertheless by no means negligible. The

206. The Black Country near Bilston, an engraving after the painting by Henry Warren (1794–1879). This picture shows well the belching chimneys and industrial blight caused by large-scale heavy industry. From *Staffordshire and Warwickshire, Past and Present* (c. 1860).

revolution of thought in the physical sciences of the eighteenth and nineteenth centuries was not of the same critical importance as that of the sixteenth and seventeenth. Indeed it might seem more proper to talk not of a revolution at all, but of an enormous spread of the results of the earlier revolution – as expressed in the Newtonian synthesis – first to other fields of science, such as heat, electricity, and chemistry, and then to the realm of economics and politics. The extensions in themselves were nevertheless in some sense radical innovations. Materially, it was through them that science first became effective in industry, that the natural forces of steam and electricity were harnessed, and that the transformation of matter, hitherto ruled by tradition, could be directed consciously to planned ends. In the field of ideas, if no comparable break with the past was made by the physical sciences, the extension to new fields brought out unsuspected aspects of Nature in the interaction of electricity and magnetism, and in the character of chemical reactions, and led to some grand generalizations such as the laws of conservation of mass and of energy and the electromagnetic theory of light.

EVOLUTION AS A SOCIAL FORCE

The really radical innovations were, however, to be furnished rather by the developments of the descriptive sciences, where mathematical analysis could still find no foothold, culminating in the great Darwinian synthesis of evolution through natural selection. Darwin's own contribution came later, as the finally inescapable conclusion of long years of geological and biological observation. It would have been acceptable long before, but for the resistance of clerical and landed interests, who felt instinctively that its acceptance meant the end of any justification of a divine ordering of the world. Newton had, through his new framework of the celestial world, largely restored the credibility of design which had been so shaken by Copernicus and Galileo; Darwin struck closer home at humanity itself. As an innovator Darwin was, quite justly, compared to Copernicus. The world of religion had survived and had, indeed, almost forgotten the upset that had been caused by the break-up of the astronomical world-picture of the ancient East. But it still had the picture of creation untouched, particularly that of man himself in the image of God; whereas after Darwin there was very little left of the book of Genesis as a literal account of history. It was some time before the appropriate face-saving formulae could be discovered, and religious truth found to be on another plane and not liable to any contradiction from vulgar facts. The suggestion that God, in his wisdom, had buried the fossils in the rocks to tempt free-thinking geologists into perdition, put forward seriously by Philip Henry Gosse, Edmund Gosse's father, was considered too far-fetched as an explanation to provide valid escape. However, as Pope Pius XII stated *ex cathedra* in 1948 that the first chapter of Genesis must be understood in an allegorical sense, the controversy must now be deemed to be over except for some Protestant fundamentalists.

The *Origin of Species* arrived at a time when its message was badly needed. It was taken up by the radical, anti-clerical wing in economics and politics, made as it was very largely in the image of its own theories of *laisser-faire* and self-help. It made possible the justification of everything that was going on in the capitalist world, the ruthless exploitation of man by man, the conquest of inferior by superior peoples. Even war itself could be justified by comparison with Nature, 'red in tooth and claw'. The old excuse for the dominance of classes or races, that they were chosen people or the sons of Gods, had faded, and new excuses were needed to justify their continuation in a rational and scientific world. Darwinism provided it, although this was the last thing Darwin himself wanted.

The fundamental importance of the theory of evolution was that it introduced a historical element into the field of science, thus breaking definitely with the orthodox branch of the Greek tradition, with the eternal truths and fixed species of Plato and Aristotle, and returning to the earlier and heretical branch of the old Ionian philosophers and of Democritus, with their emphasis on rational development and change. By bringing history into science Darwinian evolution might have been the bridge between natural and humane studies, but this it failed to be because of the strong reluctance of most of its proponents to push its doctrines home. Indeed, in its stress on the kinship of man and the animals, the social evolution of humanity was obscured in favour of a purely biological one, which was in turn to lead to the absurdity of the Nietzschean superman and to the justification of race theories and imperialism.

The links between the natural sciences and the social sciences, and the full implications of history in Nature and law in society, were not to be forged as a direct consequence of the theory of organic evolution. This was to be the work of quite another movement, at once of ideas and action, that arose as a consequence of the social effect of the Industrial Revolution and to which Marx and Engels were to give a theory and a programme. Though this occurred in the mid nineteenth century, well before the Darwinian controversy, its full meaning and consequences' were not to be apparent till the twentieth century, and the discussion of it has been left to Chapters 12 and 13.

THE SOCIAL POSITION OF THE SCIENTIST

The transition from science as a liberating idea, glimpsed at by a few choice spirits at the beginning of the eighteenth century, to a material force capable of changing the pattern of life, as it appeared to everyone at the end of the nineteenth, is not, as we have seen, one simple process, but the outcome of a conflict with many phases of alternating, rapid, or retarded advances.

In that struggle the individual scientists could not avoid being forced to consider not only the eternal order of Nature but also the consequences of successful interference with it by the new forces of technology and science. They were inevitably torn by conflicting impulses. Drawn, as most of them were, from the middle and upper classes – for the main body were easily able to assimilate and convert such individual recruits from the working classes as Faraday – they were associated with the great movements of capitalist development. Nevertheless as scientists they could not but see that the results of their efforts were being used increasingly for private enrichment, and were not leading to the improvement of the general lot of man. Only a very few scientists took a conscious part in denouncing these developments. Such were A. R. Wallace (1823–1913) and H. G. Wells (1866–1946) in Britain, Haeckel in Germany, and the group of *intellectuals* who rallied in defence of Dreyfus in France in 1894.

THE IDEAL OF PURE SCIENCE: COSMIC PESSIMISM

The majority of scientists, however, turned away from the unpleasant choice presented to them, and took refuge in a concern with the pure truths of science. They felt that if they personally were not making money out of their discoveries they were in some way free from the blame of being associated with their use for private profit.

This attitude could not fail to colour their ideas and theories even in science itself. In spite of the enormous success which scientific ideas had had in revealing the structure of the world, from the nebulae to the human brain, and in spite of the grandiose picture which the theory of evolution offered of a continuous progress, the long-range scientific outlook became, by the end of this period, essentially pessimistic. The picture of the universe was unlighted by any conception of a humanity deliberately setting itself to master Nature for the benefit of its own and subsequent generations. It therefore tended to be one of a blind fate, leading through iron laws to inescapable death.

THE LIMITS OF SCIENCE

Science appeared finite. The increasingly coherent and unitary picture of the sciences that their progress in the nineteenth century had revealed seemed to the scientists a sign that science was nearing its end. In physics, the originally separate forces, light, electricity and magnetism, and heat, were all joined together in one grand electromagnetic theory. Although gravity was not understood, its agency was fully predictable, and in fact the view of Laplace, that the whole of the universe consisted of particles whose motion would be known for all eternity if it were known at one moment, justified a picture of fate more all-inclusive than any the Greeks had had. In chemistry the elements had nearly all been discovered. Mendeleev's great generalization had even shown how many of them there could be, and how few were still to be found. In biology the Darwinian theory had shown that evolution itself had become a fatalistic progress of chance and struggle.

Of course there was still very much for science to do; each scientist in his own field saw an unlimited future of detailed discovery before him, for, oddly enough, in spite of these great generalizations of theory, science had become more specialized at the end of the nineteenth century than it had ever been before or was to remain after. Specialization itself was a way of escaping the too heavy burdens of a general view of the universe. Cosmic pessimism was balanced by confidence, if not complacency, about the present state and immediate prospects of science and society.

Whatever they felt about their own subject, nineteenth-century scientists knew that the general framework of scientific theory was secure, that the heritage of Newton had been largely fulfilled, and that the odd phenomena which did not seem to fit with this classical picture would no doubt turn out to be explainable if only someone with sufficient ingenuity would tackle them. In exactly the same way they agreed with the sentiments of the people among whom they mixed who felt that the order of society – the stock exchanges, the freedom of enterprise, the freedom of travel and trade – was, if not absolutely realized now, on the point of being realized, and that an era of indefinite intellectual and material progress was at hand. There were, of course, clouds on the horizon: labour troubles, an unpleasant increase in general armaments; but with good sense, and a realization that it was to the advantage of everyone to maintain a peaceful capitalist economy, they hoped the clouds would pass away. The future, they felt, must needs be a magnified but rather uninteresting prolongation of the past. These expectations, both in science and in society, were doomed to be disappointed in a way that we now know only too well. The twentieth century, as we shall see in the ensuing chapters of this book, was to open great and new perspectives for science and society.

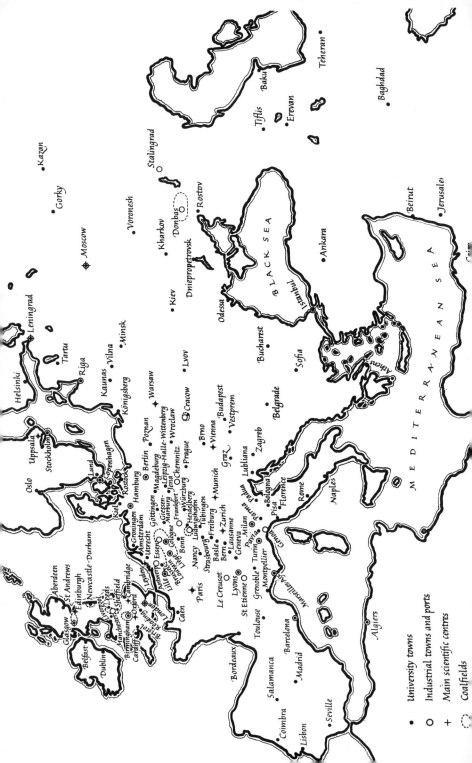

University towns

Industrial towns and ports

Main scientific centres

Coalfields

Map 4

Scientific and Industrial Europe

To illustrate the distribution of scientific and industrial centres in Europe and adjacent countries in the eighteenth, nineteenth and twentieth centuries, Chapters 8, 9, 10. Only the major industrial towns and ports are marked. The universities shown are of very different importance as centres of science; a few of the major ones are specially marked. The older foundations cluster round the central spine of Europe, shown in Map 3. In the nineteenth and even more in the twentieth century the spread to the east is marked. For the corresponding distribution of scientific centre, in America, see the insert on Map 5 (p. 1314).

Universities

Aberdeen 1494
Algiers 1879
Amsterdam 1632
Ankara 1925
Athens 1837
Baghdad 1958
Baku 1920
Barcelona 1450
Basle 1460
Beirut 1846
Belfast 1845
Belgrade 1863
Berlin 1809
Bern 1834
Birmingham 1900
Bologna 1160
Bonn 1818
Bordeaux 1441
Bristol 1909
Brno 1919
Brussels-Louvain
    1834, 1425
Bucharest 1864
Budapest 1635
Caen 1492
Cairo 970, 1908
Cambridge 1209
Cardiff 1893
Coimbra 1290
Cologne 1388
Copenhagen 1479
Cracow 1364
Dniepropetrovsk
    1918

Dublin 1591
Edinburgh 1583
Erevan 1920
Florence 1321
Frankfort 1914
Freiburg 1457
Geissen-Marburg
    1607, 1527
Geneva 1559
Glasgow 1451
Göttingen 1737
Graz 1586
Grenoble 1339
Groningen 1614
Hamburg 1919
Heidelberg 1386
Helsinki 1828
Istanbul 1883
Jena 1558
Jerusalem 1918
Kaunas 1920
Kazan 1804
Kharkov 1804
Kiel 1665
Kiev 1834
Königsberg 1544
Lausanne 1537
Leeds 1904
Leipzig-Halle-
    Wittenberg 1409,
    1694
Leningrad 1819
Leyden 1575

Liège 1817
Lille 1562
Lisbon 1911
Liverpool 1903
London 1836
Lubliana 1596
Lund 1666
Lvov 1661
Lyons 1896
Madrid 1508
Magdeburg
Manchester 1850
Marseilles-Aix 1409
Milan 1923
Minsk 1923
Montpellier 1180
Moscow 1756
Munich 1472
Nancy 1572
Naples 1224
Newcastle-Durham
    1832
Odessa 1807
Oslo 1811
Oxford 1167
Padua 1222
Paris 1160
Parma 1065
Pavia 1361
Pisa 1345
Poznan 1919
Prague 1347
Reading 1926
Riga 1862

Rome 1303
Rostock 1419
Rostov 1869
St Andrews 1411
Salamanca 1227
Seville 1502
Sheffield 1905
Sofia 1909
Stockholm 1878
Strasbourg 1567
Tartu 1632
Teheran 1935
Tiflis 1920
Toulouse 1229
Tübingen 1477
Turin 1404
Uppsala 1477
Utrecht 1636
Vestprem 1952
Vienna 1365
Vilna 1578
Voronesh 1919
Warsaw 1816
Wroclaw 1702
Würzburg 1582
Zagreb 1669
Zurich 1833

Industrial Towns and Coalfields

Antwerp
Chemnitz
Cologne

Donbas
Essen
Genoa

Le Creusot
Ludwigshafen
St Etienne

Stalingrad

Table 5 (pp. 670–71)

Science and Capitalism (Chapters 8 and 9)

In this table, which covers the eighteenth and nineteenth centuries, it is possible to give a more ordered presentation of scientific and technical progress. The first three columns cover political, intellectual, and economic developments. The central column combines the achievements of engineering and mechanics, leading up, on the one hand, to the development of heat-engines and semi-automatic machinery and, on the other, to the great central generalization of the nineteenth century, the conservation of energy and thermodynamics. The fourth column is devoted to electricity, which towards the beginning of the nineteenth century illumines chemical theory and as the twentieth approaches becomes more involved, through the telegraph and electric light, with the service of commerce and industry.

In column five we can trace the pneumatic revolution of the late eighteenth century and the more drawn-out, but equally decisive, elucidation of organic chemistry in the nineteenth, both linked at every turn with an expanding chemical industry. Finally, in the fields of Biology and Geology, we can trace the sequence between the first classification of Linnaeus and the definite establishment by Darwin of the principle of Evolution.

	Historical Events	Philosophy	Economics	Engineering and Metallurgy
— 1690				
		Locke liberty, property and toleration	Bank of England founded	
1700				*Savery* **steam pump**
	War of Spanish Succession			
	Peter the Great			*Darby* iron smelted with coke
		Berkeley idealism		*Newcomen* **steam-engine**
Chapter 8.1	Rise of Russia		Growth of small-scale manufacture in Britain and France	*Réaumur* theory of iron and steel
				Smeaton scientific engineering
		Hume scientific scepticism	Agricultural improve-ments, enclosures	
		The Philosophes		
1750	Frederick the Great	*Diderot* ' Encyclopédie' *Voltaire* **The Enlightenment**	Beginning of the Industrial Revolution	
— 1760				
	British conquest of India	*Rousseau* **'Social Contract'**		*Roebuck* Carron iron works
		Lunar society in Birmingham		*Black* latent heat
Chapter 8.2–8.4	American Revolution		*Adam Smith* ' Wealth of Nations'	*Hargreaves, Arkwright Crompton* **cotton spinning machinery**
		Kant philosophy of duty	Capitalism and the Factory System	*Boulton* metal factory *Wilkinson* ironmaster **Watt rotary engine**
	French Revolution	*Goethe* Natur-philosophie		*Cort* wrought iron
			Malthus on population	*Rumford* heat from work
— 1800				
	Napoleonic wars			*Trevithick* **high pressure engine**
				Bramah Maudslay, *Whitworth*, **machine tools**
Chapter 8.5–8.6	Holy Alliance	*Hegel* dialectical idealism	Bentham, Mill, utilitarianism	
	Peace and reaction			*Stephenson* locomotive
	Reform, triumph of bourgeoisie		Railway Age	
		Comte positivism		*Carnot* **principle of reversibility**
— 1850	Year of Revolutions	*Marx* and *Engels* **'Communist Manifesto' Dialectical materialism 'Capital'**	Britain workshop of world	*Meyer, Joule, Helmholtz* **CONSERVATION OF ENERGY** *Bessemer* **cast steel**
Chapter 8.7	American Civil War			*Lenoir* gas-engine
	Franco-Prussian War			*Siemens* **open hearth**
	Paris Commune		Great depressions	*Otto* four-stroke cycle *Gilchrist* **basic lining**
	Rise of Germany	*Mach* neopositivism	Rise of Socialism	*Clausius, Gibbs* **thermodynamics** *Parsons* turbine
	Colonial Imperialism			
— 1900				

Electricity	Chemistry	Biology and Geology
		Camerarius sex in flowers *Woodward* fossils relics of Flood
Hauksbee frictional electricity	*Stahl* phlogiston	
		Boerhaave teacher of medicine
Gray electrical conductivity *Dufay* two kinds of electricity	*Hales* begins the pneumatic revolution	*Linnaeus* classification, 'System of Nature'
Musschenbroek electric condenser and shock *Franklin* + ve. and − ve. electricity, lightning conductor	*Lomonosov* physical chemistry *Black* carbon dioxide	*Trembley* invertebrates *Buffon* 'Natural History' 'Theory of the Earth' *Haller* physiology
	Priestley, Scheele discover oxygen	
Coulomb laws	*Lavoisier* reverses phlogiston theory, founds modern chemistry	
Galvani, Volta, current electricity		*Werner* cataclysms *Hutton* geology without miracles
Davy electrochemistry	*Dalton* ATOMIC THEORY *Haüy* crystallography	*Bichat* tissues *Lamarck* evolution by modification *Oken* morphology *Cuvier* palaeontology *W. Smith* geological map
	Berzelius inorganic chemistry	*Bell, Magendie* nervous system *Baer* embryology *Lyell* 'Principles of Geology ', uniformitarianism
Oersted, Faraday, electro- magnetism The telegraph	*Dumas* *Liebig* founders of *Pasteur* organic *Kekule* chemistry *Van't Hoff*	*Schleiden and Schwann* CELL THEORY
Maxwell electromagnetic theory of light *Wilde* dynamo	Mendeleev periodic table	*Liebig, Lawes* agricultural chemistry Evidences of ice ages, and of primitive man *Mendel* heredity *Darwin* 'Origin of Species' EVOLUTION by selection *Pasteur* germ theory of disease Antisepsis, immunization
Edison electric light *Hertz* radio waves	Manufacture of dyes and explosives	

Notes

For explanation see page 369

PAGE 380. *A little studied change in agricultural methods – the growing of fodder crops like lucerne, and the consequent keeping of far more livestock through the winter – was possibly the main material factor which made the Renaissance economically successful. It started, possibly imported from the East in the fourteenth century, in Lombardy, where it was combined with irrigation mainly on country estates of wealthy town merchants. When these methods were transferred to the Low countries in the sixteenth century in an improved form they were more definitely capitalistic. Ultimately in the seventeenth and eighteenth centuries when they reached Britain they were to form the basis of the agricultural revolution which was to be a necessary complement to the Industrial Revolution (pp. 474, 511).

PAGE 408. *It is now pointed out that Copernicus made few and inaccurate astronomical observations, that his whole system was in practice no better than the one he attacked, that his reasons for choosing it were mystical rather than scientific, and that he did not see the consequences of the new ideas he was putting forward. All this may be admitted, but even granting Copernicus was as poor an astronomer as Columbus was a navigator, the essential in both cases is the spirit of innovation successfully carried through and followed up. This is the real characteristic of the Renaissance and marks a decisive break with the Middle Ages.

PAGE 416. *Fragments I have since been able to unearth show Sturtevant as a many-sided but unbalanced character. We find him giving moral advice to Prince Henry and trying unsuccessfully to sell a secret weapon to King James, a curious parallel to what Napier was doing at the same time. He wrote a book *Debre Adam* in which he evolves a universal mechanical method for learning languages which he claims Adam must have used in order to name all the animals in good Hebrew in one day. Nevertheless, in its principle it is an anticipation of the electronic translating and learning machines of today.

PAGE 425. *Kepler's life and work deserve a longer treatment than I have been able to give here.[4.47;4.100] One has had to live in the troubled parts of the twentieth century to realize the difficulties which he had to face at the time of the Thirty Years' War. He had always to work in considerable isolation, especially from other scientists, and his efforts to make contact with the other great mind of the time, Galileo, were rebuffed. He was also in his own right a considerable mathematician and may be taken as one of the founders of the integral calculus.

PAGE 461. *Blaise Pascal (1623–62) was the son of a high public official who associated with the learned men of the time.[4.90;4.110] The young Pascal showed a marked precocity, first in mathematics, where he discovered the so-called Pascal's theorem on hexagons inscribed in conic sections, fundamental to projective geometry. Later he turned to physics and ordered the carrying out of the decisive experiment on the effect of altitude on the barometer on the Puy de Dôme in his native Auvergne. He could not conduct it himself, because there was not a building of sufficient height in Paris.

To help him further in his calculation of taxes, he invented and actually made a number of calculating machines – the lineal ancestors of those of today. Despite his mathematical and philosophical distinction, Pascal was a gifted *entrepreneur* and business man. He organized the factory production of his calculating machines and took great pains to have them advertised. In 1662 he founded in Paris the first omnibus company in the world (the profits were to be given to charity) and he tried simultaneously to found one in London and in Amsterdam.

A great change came about in his life in 1654 when he became associated with the Jansenists, an effectively puritan sect of Catholics who were being persecuted by the more flexible and dominant order of Jesuits, and undertook the defence of the Jansenists in his celebrated *Lettres provinciales* (1656–7). After his conversion he withdrew more and more from science and society, and gave himself over to philosophical speculations and to religion. He had already laid the foundations of the theory of probability, from his study of gaming, and used it to propound 'Pascal's wager', that however improbable the truths of religion were, they should be believed because the reward of eternity in heaven would always be sufficient to recompense the believer.

For the rest of his short life, he lived as an ascetic and contemplative recluse. But his profound *Pensées*, which were published after his death, have remained one of the glories of French philosophy and letters.

PAGE 480. *John Harrison and his brother James were typical clock-makers from Lincolnshire. They showed great ingenuity not only in design but also in the choice of materials. This enabled them to solve the problem that had baffled all clock-makers at a time when the clock-making art was one of the finest branches of mechanics and also a gentleman's hobby. The ill-fated Louis XVI was an amateur clock-maker. The Harrisons found that making a clock that gave accurate time to within a few seconds a day was simple compared with the task of extracting money from the Admiralty. Indeed, John Harrison would not even have done this had he not secured the support of the King, George III, that much maligned monarch with a weakness for science. His patronage was effective not only in this case but in setting up an Observatory of his own in which his former music master, Herschel, was able to discover the new planet Uranus and sketch the first idea of the galaxy.

PAGE 488. *The religious importance of Newton's world view was not overlooked by his followers or his critics. A violent controversy was carried on between Leibniz and the Newtonians from 1705 to 1716, culminating in the published correspondence between Samuel Clarke, D.D.(1675–1729) and Leibniz.[4.25] Clarke, an ardent follower of Newton and liberal theologian, was, according to Voltaire, only prevented from becoming Archbishop of Canterbury by a remark from the Bishop of Lincoln to Princess Caroline that Clarke was the most learned and honest man in her dominions but had one defect – he was not a Christian. The dispute is essentially the charge that Newton's work contributed to the decline of natural religion, mainly on account of his view that space was the sensorium of God. This argument had a double interest, theologico–political and scientific–philosophical. Both sides were trying to prove that their own views were harmless to the Establishment, but Leibniz smelled out the essentially revolutionary character of the Newtonian philosophy and opposed it for the very reason that Voltaire embraced it. Scientifically the argument turned on space and time and the nature of occult forces such as gravity. Here Leibniz had the worst of it – for the time – but his critical arguments were part of the tradition that inspired Einstein's supersession of Newtonian theory.

PAGE 489. *Professor Dingle uses this statement to confute my thesis of the parallelism of economics and ideology, which he attacks as being at the same time vague and incorrect. If, he asks, the establishment by Newton of a *dynamic* order instead of the Greek–Medieval *static* one corresponded to the birth of capitalism, what must correspond to Einstein's return to static conceptions in the twentieth century? According to him: 'We ought therefore to have in these days a reversion to a fixed hierarchical order in which man knows his place.' Now here I must thank Professor Dingle for presenting me with a parallel which had escaped me. The essence of the twentieth-century scientific advance is the notion of implicit rather than explicit functions in the mathematical sense, or in common language the way in which everything is related to everything else. Societies in the past have been centrally directed or have operated as the result of a large number of independent market deals between individuals. The very nature of the modern industrial State is one of complex inter-related organization which will impose in the long run a socialist pattern, and this corresponds to the new character of science. Now Professor Dingle calls these correspondences mystical. By this he means that they cannot be expressed in precise definitions and propositions. In this view all history, beyond mere chronicle, and all social science is mystical. I would contend, on the other hand, that they have a real meaning, that general modes of thinking are transferable from the scientific to the social field, and vice versa. The value of stating them is to make one aware of what one is doing or thinking.

PAGE 514. *The object of the German academies was somewhat more practical than would seem to be implied by these remarks. The study of natural resources, particularly minerals, appears in all their programmes. Leibniz himself, aiming at a national and linguistic revival of Germany, certainly did not scorn them. He even foresaw the main line of advance when he stated that 'Germans have always excelled in mining, Germany should therefore become the mother of chemistry.' All through the eighteenth century, in and around the academies, noblemen and aspiring bureaucrats occupied themselves in improvements in agriculture and industry. This was not seen as a bar to artistic expression. One of Goethe's loveliest lyrics, 'Über allen Gipfeln ist Ruh', was composed during a tour of inspection of mines.

PAGE 520. *It seems to be a general rule that in its initial stages every new product or process that succeeds in meeting a need increases by the exponential law, in which the rate of growth is proportional to the size, so that, for instance, it doubles every so many years. A useful criterion for a revolution is when this period of doubling is shorter than a human generation. This certainly applies to the commodities coal, cotton, iron, and steam-power in the Industrial Revolution, the average periods of doubling being 20, 6, 10, 16 years respectively. Slower rates of growth produce changes that are noticed only by historians. Naturally no rate of growth, certainly not a rapid one, can increase indefinitely. Normally as saturation is reached the rate first becomes steady then tapers off to nothing. However, either in a declining economy or in an expanding one, when new products are being rapidly introduced one after the other, decline and extinction is the final state. As Price has shown, the exponential rule applies very exactly to knowledge itself, measured by the number of papers published; the present scientific revolution being marked by a doubling every ten years.[5.93] Price's conclusion that this cannot go on until everyone is a scientist is certainly true, but it still has a long way to go before there is no one left to read the papers.

PAGE 523. *Of the remaining industrial areas the West Riding, solidly based on the older cloth and cutlery, came somewhat later into the movement of the Industrial Revolution.[5.31] South Wales linked with the tin and copper of Cornwall and with the Severn area, including the classical Coalbrookdale of the Darbys, were important secondary centres. especially for the development of the steam engine and iron metallurgy. [5.2;5.8]

PAGE 526. *The special importance of Birmingham in the eighteenth century arose just from the fact it was not like one of the older towns with their guild structure. Apart from this it attracted the dissenters, who were mainly engaged in small-scale metal working, for which they could draw on the coal and iron of the neighbouring Black Country. It was possibly because it was lacking in all natural advantages, including a navigable river, that Birmingham put such demands on the ingenuity of its inhabitants.

PAGE 531. *An early attempt to work out the consequences of interatomic forces was made by Roger Boscovich (1711–78), a Serbian Jesuit priest who taught mathematics at Rome.[4.37;4.169] He postulated effectively point atoms surrounded by force fields which were repulsive at very small distances, attractive at very large distances, and might be alternatively attractive and repulsive at intermediate distances. This is precisely the form of interatomic forces as we know them today, and we can even use some of Boscovich's expressions for them. However, although his abilities were recognized, his theories were still too complex to deal with the data available for many years after his death and he had little effect on the general course of science.

PAGE 540. *The English poet Blake at this time identified the darker side of the Industrial Revolution, 'The dark Satanic mills', with the application of the philosophy of Newton, and reacting to both found that cold reason was pointing the way to Hell.

> I turn my eyes to the Schools and Universities of Europe
> And there behold the Loom of Locke, whose Woof rages dire,
> Wash'd by the Water-wheels of Newton: black the cloth
> In heavy wreathes folds over every Nation: cruel Works
> Of many Wheels I view, wheel without wheel, with cogs tyrannic
> Moving by compulsion each other, not as those in Eden, which
> Wheel within Wheel, in freedom revolve in harmony and peace.

PAGE 558. *Alfred Russel Wallace, the independent co-discoverer of natural selection, was nevertheless the first eminent scientist who adhered to socialism (p. 664).

PAGE 565. *Dr Richter of the Hahn-Meitner Institute, Berlin, complains that I have been unfair in my judgement of the old medicine. I would still stand by my remarks here about its inefficacy, but other things that I did not mention might still be taken into consideration. The study of medicine has the longest and most continuous role in the history of science. (I have referred to it in this book in more than a dozen places.) In its earliest phase, with a shaman or medicine man, we have the universal wise man who controls the heavens, health and disease, and the fortunes of hunting and agriculture. From him are descended not only doctors and astronomers, but also poets and rulers. All through history the doctor has remained as one whose value to society lies in his learning – the first effectively subsidized scientist. But this has not been an unmixed blessing. The closeness of the link between medicine and astronomy, particularly in Greek times, was to introduce the wholly fictitious idea of order from the heavens to man – the macrocosm and the microcosm. The practical experience of doctors was here forced into the strait-jacket of Aristotelean physics of the elements or the Chinese Yang and Yin theories. Medical theory shows itself, in retrospect, far more cramping to the advance of medicine than theology was to astronomy. But the physician had an essential corrective in his contact with actual people and their diseases. The doctor had an accumulated experience, and in his clinical observations and diagnoses, at least from Hippocrates onwards, he could certainly offer to calm the fears of his patients, even if he could do little to cure them. From the Middle Ages the profession of medicine was established as one of the professional faculties of the University – medicine, law and theology – for doctors, lawyers and clergymen. If science had little effect on the development of medical theory and practice, the doctors made up for it by their contribution as individuals to the growth of chemical and biological science. It was only at the end of the nineteenth century and throughout the present century that a juncture between clinical and scientific medicine has been really effected and the age of scientific medicine inaugurated.

PAGE 567. *Japanese militarism was to find its nemesis in the Second World War, but subsequent events have shown the surviving power of the science of the Japanese and their capacity, when they have freed themselves from foreign occupation, to make a great contribution to non-European science, indeed to world science.

PAGE 582. *Apparently Watt only made use of the governor; he found it already in action, invented by some unknown millwright as a regulator of the stones of water-mills.[5,41] The centrifugal governor is not strictly speaking the first cybernetic device. The regulation of such delicate mechanisms as clocks had been achieved long before. It was the first of such devices, however, to be used to control practical mechanical power.

PAGE 588. *The idea of equivalence of different forms of energy (or force) and of their conservation is much older. It was indeed formulated by Leibniz in 1715 in the first of his letters attacking the Newtonian philosophy. 'According to my opinion the same force and vigour remains always in the world and only passes from one part of matter to another agreeably to the laws of Nature and the beautiful pre-established order.'[4][,25] This doctrine, however, put forward in an essentially theological argument and without the steam-engine to give it any practical relevance, was to remain dormant for over a century.

It has been pointed out to me by Dr Richter that the conservation of energy as such, not its transformation, is included in the golden rule of mechanics, but what I am insisting on here is the transformation of energy from one kind to another, a universal *currency* of energy.[5][,3]

PAGE 594. *The principle of the slide-rest and screw-cutting lathes was not new. What the mechanics of the early nineteenth century were doing was to rediscover and adapt to hard and useful work on steel what the artistic and ingenious wood, ivory, and brass turners of the sixteenth and seventeenth centuries had already done. There is still extant in the Hermitage at Leningrad a complete set of magnificently made semi-automatic and copying lathes used by Peter the Great mostly for making medals and candelabra.

PAGE 603. *The idea of torsion and suspension was in fact used by the Chinese as far back as the T'ang dynasty. But Coulomb came to it rather from the opposite direction, by studying the strength of materials for the French Admiralty and measuring the amount of torsion that a steel bar could stand as an index to its quality.

PAGE 620. *Dr Richter has objected to this comparison, and written that it is stretching the principles of the history of science to compare the phlogiston with the electron theory of valency. I would point out that my intention was to make a purely formal comparison, in that the phlogiston theory brought together a whole series of phenomena in which combustion seemed to imply a loss of something rather than a gain. Although the idea is very crude, wood turns into ashes, lead to dross, and iron to rust – all these seem to represent the same type of phenomenon. They call for a unitary explanation, which is given by the phlogiston theory. I had not intended to imply that the phlogistonists had grasped anything about the nature of electrons. However, their unitary theory enabled the phlogistonists to think in a coherent way about a great variety of chemical processes and even to carry out systematic experimental studies of the new elements as they appeared. Dr Richter's argument could be used equally well to claim that the Greeks had no idea of the real atoms, as they were discovered afterwards; but their abstract atomism, philosophic as much as material, was an essential step in the discovery or rather the rediscovery of atomism in the seventeenth century.

PAGE 638. *Linnaeus was very fond of this pretty little northern flower. When ennobled in 1761, he included it in his coat-of-arms, and held it in all portraits showing his hands.

PAGE 652. *This reference to Claude Bernard is altogether inadequate. Though he made less impact in his time, as his physiological discoveries were not of immediate application, they were to be the basis of all modern physiology and biochemistry, of the work of Pavlov as well as Hopkins. More than that, his *Introduction to the Study of Experimental Medicine*[5][.22] remains an almost unique example of an analysis of the processes of science by one who was a great and creative scientist rather than a philosopher. As time goes on his influence on science is bound to grow.

PAGE 654. *Here the mechanical ingenuity was far ahead of any scientific knowledge of the growth of plants. The enormous success of agricultural machinery was due rather to its labour-saving capacity than to any increase in harvest yield. In medieval or earlier agriculture the growing of crops was essentially on a subsistence basis, as it still is over two-thirds of the world. To win the harvest, 'hands' had to be available with sickles and women to bind the sheaves, in the greatest number possible. These had to live through the rest of the year as best they could. Overall, the operations of farming ate up most of the produce. Even in good years, the peasants went short part of the time between harvests; in bad years they starved.

Effectively, non-mechanical agriculture gives a surplus of between five and ten per cent, which is needed to feed the towns. Mechanical agriculture, on the other hand, can provide not necessarily such good yields but can provide them with far less labour. Because of that the agricultural population in the main grain-producing regions can be reduced to some five per cent. With greater knowledge, the idea of machines simply reproducing the age-old manual or horse-drawn agricultural machinery will be drastically revised. The very word 'agriculture', which depends on the *ager*, the plough (p. 115), is obsolete. Ploughs will be replaced by chemicals applied to undisturbed soil, and the energy required to carry out agricultural procedures may be very much reduced at the same time as giving much bigger yields.

Bibliography to Volume 2

PART 4

1. BUTTERFIELD, H., *The Origins of Modern Science*, 2nd ed., London, 1957
2. BOAS, M., *The Scientific Renaissance, 1450–1630*, London, 1962
3. CLARK, G. N., *Science and Social Welfare in the Age of Newton*, 2nd ed., Oxford, 1949
4. CLARK, G. N., *The Seventeenth Century*, Oxford, 1929
5. DOBB, M., *Studies in the Development of Capitalism*, London, 1946
6. HADYN, H., *The Counter-Renaissance*, New York, 1950
7. HALL, A. R., *The Scientific Revolution*, 2nd ed., London, 1962
8. HARVEY, W., *The Anatomical Exercises of William Harvey: De Motu Cordis 1628; De Circulatione Sanguinis 1649*, London, 1949
9. HUTTEN, E. H., *The Origins of Science, an Inquiry into the Foundations of Western Thought*, London, 1962
10. KOYRÉ, A., *La Révolution Astronomique*, Paris, 1961
11. KUHN, T. S., *The Copernican Revolution: Planetary Astronomy in the Development of Western Thought*, London, 1957
12. KUHN, T. S., *The Structure of Scientific Revolutions*, Chicago, 1962
13. LYONS, Sir H., *The Royal Society 1660–1940*, Cambridge, 1944
14. NEF, J. U., 'The Genesis of Industrialism and of Modern Science (1560–1640)', *Essays in Honour of Conyers Read*, ed. N. Downs, Chicago, 1953
15. ORNSTEIN, M., *The Role of Scientific Societies in the Seventeenth Century*, Chicago, 1938
16. PLEDGE, H. T., *Science Since 1500*, London, 1940
17. SARTON, G., *The Appreciation of Ancient and Medieval Science During the Renaissance*, London, 1956
18. SMITH, P., *A History of Modern Culture*, 2 vols., London, 1930, 1940
19. WELD, C. R., *A History of the Royal Society*, 2 vols., London, 1848
20. WIGHTMAN, W. P. D., *Science and the Renaissance*, 2 vols., Edinburgh, 1962
21. WILLEY, B., *The Seventeenth Century Background*, London, 1934
22. WOLF, A., *A History of Science, Technology, and Philosophy in the Sixteenth and Seventeenth Centuries*, 2nd ed., London, 1950

23. AGRICOLA, G., *De Re Metallica*, trans. H. C. and L. H. Hoover, London, 1912
24. ALBERTI, L. B., *Trattato della Pittura*, Milan, 1804
25. ALEXANDER, H. G., *The Leibniz–Clarke Correspondence*, Manchester, 1956
26. ANDRADE, E. N. DA C., *A Brief History of the Royal Society*, London, 1960
27. ANDRADE, E. N. DA C., *Isaac Newton*, London, 1950
28. ANTAL, F., *Florentine Painting and its Social Background*, London, 1947

29. ARMITAGE, A., *Copernicus the Founder of Modern Astronomy*, London, 1938
30. BACON, F., *The Works of Francis Bacon*, ed. J. Spedding, R. L. Ellis, and D. D. Heath, 14 vols., London, 1857–74
31. BELL, A. E., *Christian Huygens and the Development of Science in the Seventeenth Century*, London, 1947
32. BELON, P., *L'Histoire de la Nature des Oyseaux*, 7 vols., Paris, 1553
33. BERNAL, J. D., *The Social Function of Science*, London, 1939
34. BERNAL, J. D., 'Leonardo da Vinci', *Labour Monthly*, vol. 34, 1952
35. BISHOP, W. J., *The Early History of Surgery*, London, 1960
36. BOAS, M., *Robert Boyle and Seventeenth Century Chemistry*, Cambridge, 1958
37. BOSCOVICH, R. J., *A Theory of Natural Philosophy*, London, 1922
38. BRECHT, B., *The Life of Galileo*, London, 1960
39. BOURNE, W., *A Regiment for the Sea*, London, 1592
40. BOYLE, Hon. R., *The Works of the Honourable Robert Boyle*, London, 1744
41. BROWN, H., *Scientific Organizations in Seventeenth-Century France (1620–1680)*, Baltimore, 1934
42. BULLEN, A. H., *Elizabethans*, London, 1924
43. BURCKHARDT, J., *The Civilization of the Renaissance in Italy*, London, 1944
44. BURTT, E. A., *The Metaphysical Foundations of Modern Physical Science*, London, 1925
45. BUSH, J. N. D., *Science and English Poetry*, Oxford, 1950
46. CARSWELL, J., *The South Sea Bubble*, London, 1960
47. CASPAR, M., *Kepler*, trans. and ed. C. D. Helman, London, 1959
48. CHAMBERS, R. W., *Thomas More*, London, 1935
49. CLARK, Sir G. N., *War and Society in the Seventeenth Century*, Cambridge, 1958
50. COHEN, I. B., 'Neglected Sources for the Life of Stephen Gray (1666 or 1667–1735)', *Isis*, vol. 45, 1954
51. CREW, H., *The Rise of Modern Physics*, Baltimore, 1928
52. CROMBIE, A. C., *Robert Grosseteste*, Oxford, 1953
53. CROWTHER, J. G., *Founders of British Science: Wilkins, Hooke, Boyle, Ray, Wren, Newton*, London, 1960
54. CROWTHER, J. G., *Francis Bacon. The First Statesman of Science*, London, 1960
55. CROWTHER, J. G., *Six Great Astronomers*, London, 1961
56. DEBENHAM, F., *Discovery and Exploration. An Atlas–History of Man's Journeys into the Unknown*, London, 1960
57. DE SANTILLANA, G., *The Crime of Galileo*, London, 1958
58. DESCARTES, R., *Discourse on Method* (Everyman), London, 1949
59. DESCARTES, R., *Geometria*, Leyden, 1649
60. DIBNER, B., *Leonardo da Vinci*, New York, 1946
61. DUGAS, R., *Histoire de la Mécanique*, Neuchâtel, 1950
62. ERNOUF, A. A., *Denis Papin*, Paris, 1874
63. 'ESPINASSE, M., *Robert Hooke*, London, 1956
64. FAHIE, J. J., *Galileo: His Life and Works*, London, 1903
65. FARRINGTON, B., *Francis Bacon, Philosopher of Industrial Science*, London, 1951
66. FELDHAUS, F. M., *Leonardo: der Techniker und Erfinder*, Jena, 1913
67. FOSTER, Sir M., *Lectures in the History of Physiology during the Sixteenth, Seventeenth, and Eighteenth Centuries*, Cambridge, 1901
68. FRANKLIN, K. J., *William Harvey, Englishman, 1578–1657*, London, 1961
69. GALILEI, G., *Dialogue Concerning the Two Chief World Systems – Ptolemaic and Copernican*, trans. S. Drake, London, 1962

70. GALILEI, G., *Dialogues Concerning Two New Sciences*, trans. H. Crew and A. de Salvio, New York, 1914
71. GALILEI, G., 'The Sidereal Messenger', extracts in *The Autobiography of Science*, ed. F. R. Moulton and J. J. Schifferes, 2nd ed., London, 1963
72. GALILEI, G., *The System of the World*, trans. T. Salusbury, London, 1661
73. GASSENDI, P., *Opera Omnia*, 6 vols., Lyons, 1658
74. GESNER, C., *Historiae Animalium*, 4 vols., Zurich, 1551–8
75. GILBERT, W., *De Magnete*, ed. S. P. Thompson, London, 1901
76. GILLISPIE, C. C., *The Edge of Objectivity: an Essay in the History of Scientific Ideas*, Oxford, 1960
77. GINSBERG, M., *The Idea of Progress*, London, 1953
78. GLANVILL, J., *Plus Ultra*, London, 1668
79. GRESHAM COLLEGE, *An Account of Rise, etc., of Gresham College*, London, 1707
80. HAKLUYT SOCIETY, *A Regiment for the Sea and Other Writings on Navigation*, by W. Bourne, ed. E. G. R. Taylor, Cambridge, 1963
81. HAKLUYT SOCIETY, *Select Documents Illustrating the Four Voyages of Columbus*, ed. C. Jane, series II, vol. 65, London, 1930
82. HALL, A. R., *Ballistics in the Seventeenth Century*, London, 1952
83. HARTLEY, H. (ed.), *The Royal Society – Its Origins and Founders*, London, 1960
84. HARVEY, G., *Letter-Book of Gabriel Harvey*, ed. E. J. L. Scott, London, 1884
85. HILL, C., and DELL, E., *The Good Old Cause: The English Revolution of 1640–60*, London, 1949
86. HILL, J., *Review of the Work of the Royal Society of London*, London, 1760
87. HOBBES, T., *Leviathan*, ed. M. Oakeshott, Oxford, 1946
88. HOOKE, R., *The Diary of Robert Hooke*, ed. H. W. Robinson and W. Adams, London, 1935
89. HOOKE, R., *Micrographia*, London, 1665
90. HUMBERT, P., *L'Oeuvre Scientifique de Blaise Pascal*, Paris, 1947
91. HUYGENS, C., *Oeuvres Complètes de Christiaan Huygens*, 22 vols., The Hague, 1888–1950
92. JOHNSON, F. R., *Astronomical Thought in Renaissance England*, Baltimore, 1937
93. JOHNSON, F. R., 'Gresham College', *Journal of the History of Ideas*, vol. I, 1940
94. JOHNSON, S., *Lives of the English Poets*, 2 vols., London, 1906
95. JONES, R. F., *Ancients and Moderns: A Study of the Rise of the Scientific Movement in Seventeenth Century England*, 2nd ed., Washington, 1961
96. JONES, R. F., *The Seventeenth Century*, Stanford, California, 1951
97. JONES, R. F., *The Triumph of English Language*, London, 1953
98. KING, H. C., *The History of the Telescope*, London, 1955
99. KEPLER, J., *Opera Omnia*, ed. C. Frisch, Frankfurt, 1858–71
100. KOESTLER, A., *The Sleepwalkers*, London, 1959
101. KOYRÉ, A., *Études Galiliennes*, Paris, 1939
102. LOVEJOY, A., *The Great Chain of Being*, Cambridge, Mass., 1936
103. MCCURDY, E., *The Mind of Leonardo da Vinci*, London, 1932
104. MASSON, F., *Robert Boyle: A Biography*, London, 1914
105. MAYOW, J., *Medico-Physical Works*, Oxford, 1926
106. MERSENNE, M., *Correspondence du M. Mersenne*, ed. C. de Waard, 3 vols., Paris, 1945–6
107. MERTON, R. K., *Science, Technology, and Society in the Seventeenth Century*, Bruges, 1938

108. MEYER, R. W., *Leibnitz and the 17th Century Revolution*, trans. J. P. Stern, Cambridge, 1952

109. MILTON, J., *Areopagitica*, ed. J. W. Hales, London, 1949

110. MORTIMER, E., *Blaise Pascal*, London, 1959

111. NEEDHAM, J. (ed.), *The Teacher of Nations: Comenius*, Cambridge, 1942

112. NEF, J. U., *The Rise of the British Coal Industry*, 2 vols., London, 1932

113. NEWTON, I., *The Correspondence of Isaac Newton*, ed. H. W. Turnbull, 3 vols., Cambridge, 1960–61

114. NEWTON, I., *The Mathematical Principles of Natural Philosophy* (Motte's translation revised by F. Cajori), Berkeley, California, 1947

115. NEWTON, I., *Opticks*, London, 1704

116. NICOLSON, M. H., *The Breaking of the Circle*, Evanston, Illinois, 1950

117. NORMAN, R., *The Newe Attractive*, London, 1581

118. OLSCHKI, L., *Genius of Italy*, London, 1950

119. PAGEL, W., 'A Background Study to Harvey', *Medical Bookman and Historian*, vol. 2, p. 407

120. PATTERSON, L. D., 'Hooke's Gravitation Theory and its Influence on Newton', parts I and II, *Isis*, vols. 40 and 41, 1949 and 1950

121. PELISSON-FONTANIER, P., *The History of the French Academy*, trans. H. S., London, 1657

122. PILKINGTON, R., *Robert Boyle: Father of Chemistry*, London, 1959

123. POLVANI, G., 'L'Invention de la Pile', *Revue d'Histoire de Sciences*, vol. 2, 1949

124. PURVER, M., and BOWEN, E. J., *The Beginning of the Royal Society*, Oxford, 1960

125. RABELAIS, F., *Works*, Navarre Society, London, 1948

126. RANDALL, J. H., *The School of Padua and the Emergence of Modern Science*, Padova, 1961

127. REY, J., *The Essays of Jean Rey*, ed. D. McKie, London, 1951

128. ROYAL ASTRONOMICAL SOCIETY, 'Nicolaus Copernicus, De Revolutionibus, Preface and Book I' (trans. J. P. Dobson, assisted by S. Brodetsky), *Occasional Notes*, no. 10, May, 1947

129. ROYAL SOCIETY, *Newton Tercentenary Celebrations*, Cambridge, 1947

130. SARTON, G., *Six Wings: Men of Science in the Renaissance*, London, 1958

131. SCIENCE AT THE CROSS ROADS, Papers presented to the International Congress of the History of Science and Technology, by delegates of the USSR, London, 1931

132. SHERRINGTON, Sir C. S., *The Endeavours of Jean Fernel*, Cambridge, 1946

133. SHERRINGTON, Sir C. S., *Man on His Nature*, Cambridge, 1940

134. SIGERIST, H. E., *Four Treatises of Theophrastus von Hohenheim called Paracelsus*, Baltimore, 1941

135. SINGER, C., *The Earliest Chemical Industry*, London, 1948

136. SINGER, D. W., *Giordano Bruno, His Life and Thoughts*, London, 1950

137. SMITH, C., and GNUDI, M. T., *The Pirotechnia of Vanoccio Biringuccio*, New York, 1941

138. SPINOZA, B. DE, *Tractatus Theologico-politicis*, trans. R. H. M. Elwes, London, 1895

139. SPRAT, T., *The History of the Royal Society of London*, London, 1667

140. SPRAT, T., *History of the Royal Society*, ed. W. H. Jones, J. T. Copeland, London, 1959

141. STENO, N., [Canis Carcharix Dissectum Caput], *The Earliest Geological Treatise* (*1667*), trans. A. Garb, London, 1959

142. STILLMAN, J. M., *Theophratus Bombastus von Hohenheim called Paracelsus*, Chicago, 1920
143. STIMSON, D., *Scientists and Amateurs: A History of the Royal Society*, London, 1949
144. STRAKER, E., *Wealden Iron*, London, 1931
145. STURTEVANT, S., 'Metallica' (1612), *Supplement to the Series of Letters Patent, etc. (1617–1852)*, vol. I, London, 1858
146. SYFRET, R. H., 'The Origins of the Royal Society', *Notes and Records of the Royal Society of London*, vol. 5, 1948
147. TAWNEY, R. H., *Religion and the Rise of Capitalism*, London, 1927
148. TAYLOR, E. G. R., *Late Tudor and Early Stuart Geography 1583–1650*, London, 1934
149. TAYLOR, E. G. R., *The Mathematical Practitioners of Tudor and Stuart England*, Cambridge, 1954
150. TAYLOR, E. G. R., *Tudor Geography 1485–1583*, London, 1930
151. TAYLOR, F. S., *Galileo and the Freedom of Thought*, London, 1938
152. THORNDIKE, L., *A History of Magic and Experimental Science*, 8 vols., New York, 1923–58
153. TILLYARD, E. M. W., *The Elizabethan World Picture*, London, 1943
154. TURNBULL, H. W., *The Great Mathematicians*, 4th ed. rep., London, 1962
155. TURNBULL, G. H., *Hartlib, Dury and Comenius*, Liverpool, 1847
156. TURNER, D. M., *Makers of Science, Electricity and Magnetism*, Oxford, 1926
157. UNWIN, G., *Industrial Organization in the Sixteenth and Seventeenth Centuries*, Oxford, 1904
158. VAN DEUSEN, N. C., *Telesio, the First of the Moderns*, New York, 1932
159. VAVILOV, S. I., *Isaac Newton*, Vienna, 1948
160. VESALIUS, A., *De Humani Corporis Fabrica*, Basle, 1543
161. VESPUCCI, A., *Letters of A. Vespucci*, ed. C. R. Markham, 1894
162. VICO, G. B., *The New Science of Giambattista Vico*, trans. T. G. Bergin and M. H. Fisch, New York, 1948
163. VICO, G. B., *The Autobiography of Giambattista Vico*, trans. T. G. Bergin and M. H. Fisch, New York, 1944
164. VINCI, LEONARDO DA, *The Notebooks of Leonardo da Vinci*, ed. E. McCurdy, 2 vols., London, 1938
165. VINCI, LEONARDO DA, *Paragone*, London, 1949
166. VIVES, J. L., *On Education*, trans. F. Watson, Cambridge, 1913
167. WARD, J., *The Lives of the Professors of Gresham College*, London, 1740
168. WHITTAKER, E. T., *A History of the Theories of the Ether and Electricity*, vol. I, London, 1951
169. WHYTE, L. L. (ed.), *Roger Joseph Boscovich, S.J., F.R.S., 1711–1787*, London, 1961
170. WINSTANLEY, G., *Selections from His Works*, ed. L. Hamilton, London, 1944
171. WRIGHT, L. B., *Middle-Class Culture in Elizabethan England*, Oxford, 1935

PART 5

1. ASHTON, T. S., and SYKES, J., *The Coal Industry of the Eighteenth Century*, Manchester, 1929
2. ASHTON, T. S., *Iron and Steel in the Industrial Revolution*, Manchester, 1924
3. BERNAL, J. D., *Science and Industry in the Nineteenth Century*, London, 1953
4. CLOW, A. and N., *The Chemical Revolution*, London, 1952
5. CROWTHER, J. G., *British Scientists of the Nineteenth Century*, London, 1935
6. CROWTHER, J. G., *Famous American Men of Science*, London, 1937
7. CROWTHER, J. G., *Scientists of the Industrial Revolution*, London, 1962
8. DICKINSON, H. W., *A Short History of the Steam Engine*, Cambridge, 1939
9. HOBSBAWM, E. J., *The Age of Revolution*, London, 1962
10. MANTOUX, P., *The Industrial Revolution in the Eighteenth Century*, London, 1931
11. WILLEY, B., *The Eighteenth Century Background*, London, 1940
12. WOLF, A., *A History of Science, Technology, and Philosophy in the Eighteenth Century*, 2nd ed., London, 1952

13. ACADEMY OF SCIENCES, USSR, *220 let Akademii Nauks SSR*, Moscow, 1945
14. APPERT, C., *L'Art de Conserver Pendant Plusieurs Années Toutes les Substances Animales et Végétales*, Paris, 1810
15. ARMYTAGE, W. H. G., 'The Royal Society and the Apothecaries 1660–1722', *Notes and Records of the Royal Society*, vol. II, 1954, p. 33
16. BABBAGE, C., *Passages from the Life of a Philosopher*, London, 1864
17. BABBAGE, C., *Reflections on the Decline of Science in England*, London, 1830
18. BAILEY, Sir E., *Charles Lyell*, London, 1962
19. BAINES, Sir E., *History of the Cotton Manufacture in Great Britain*, London, 1835
20. BEER, G. R. DE, *Sir Hans Sloane and the British Museum*, London, 1953
21. BERNAL, J. D., *The Freedom of Necessity*, London, 1949
22. BERNARD, C., *An Introduction to the Study of Experimental Medicine*, trans. H C. Greene, New York, 1949
23. BURNS, C. D., *A Short History of Birkbeck College*, London, 1924
24. BURSTALL, A. F., *A History of Mechanical Engineering*, London, 1963
25. CARNOT, S., *Sur la Puissance Motrice du Feu*, Paris, 1824
26. CARSWELL, J., *The Prospector*, London, 1950
27. CHEVALIER, J., *Le Creusot*, Paris, 1946
28. CRAMP, W., *Michael Faraday and Some of His Contemporaries*, London, 1931
29. CROWTHER, J. G., *Six Great Doctors*, London, 1957
30. CROWTHER, J. G., *The Story of Agriculture*, London, 1958
31. CRUMP, W. B. (ed.), *The Leeds Woollen Industry, 1780–1820*, Leeds, 1931
32. DANILEVSKII, V. V., *Russkaya Tekhnika*, Moscow, 1948
33. DARWIN, C. R., *The Effects of Cross and Self Fertilization in the Vegetable Kingdom*, London, 1876
34. DARWIN, C. R., *The Expression of the Emotions in Man and Animals*, London, 1872
35. DARWIN, C. R., *The Formation of Vegetable Mould Through the Action of Worms*, London, 1881
36. DARWIN, C. R., *A Naturalist's Voyage*, London, 1860
37. DARWIN, C. R., *The Origin of Species*, London, 1859

38. DE BEER, G., *The Sciences Were Never at War*, London, 1960
39. DICKENS, C., *Hard Times*, London, 1854
40. DICKINSON, H. W., *James Watt*, Cambridge, 1936
41. DICKINSON, H. W., *Matthew Boulton*, Cambridge, 1937
42. DICKINSON, H. W., and TITLEY, A., *Richard Trevithick*, Cambridge, 1934
43. DUBOS, R. J. W., *Louis Pasteur*, London, 1951
44. EASTWOOD, W., *Science and Literature*, London, 1957
45. ENGELS, F., *The Condition of the Working Class in England in 1844*, London, 1892
46. FARADAY, M., *Experimental Researches in Electricity*, 3 vols., London, 1855
47. FARADAY, M., *Faraday's Diary*, ed. T. Martin, 8 vols., London, 1932–6
48. GALTON, Sir F., *Hereditary Genius*, London, 1869
49. GEDDES, P., *Cities in Evolution*, London, 1915 .
50. GIEDION, S., *Mechanization Takes Command*, Oxford, 1948
51. GILLAM, J. G., *The Crucible: The Story of Joseph Priestley, LLD, FRS*, London, 1954
52. GILLISPIE, C. C., *Genesis and Geology*, Harvard, 1951
53. GOURLIE, N., *The Prince of Botanists: Carl Linnaeus*, London, 1953
54. HABAKKUK, H. J., *American and British Technology in the Nineteenth Century*, Cambridge, 1962
55. HADFIELD, Sir R. A., *Faraday and His Metallurgical Researches*, London, 1931
56. HAMMOND, J. L. and B., *The Town Labourer, 1760-1832*, London, 1917
57. HAMMOND, J. L. and B., *The Village Labourer, 1760-1832*, London, 1912
58. HART, I. B., *James Watt and the History of Steam Power*, London, 1958
59. HARTOG, Sir P., 'Joseph Priestley and his Place in the History of Science'. *Proceedings of the Royal Institution of Great Britain*, April, 1931
60. HOBSON, J. A., *The Evolution of Modern Capitalism*, London, 1930
61. HOWARTH, O. J. R., *The British Association for the Advancement of Science (1831–1931)*, London, 1931
62. HOWARTH, O. J. R. (ed.), *London and the Advancement of Science*, London, 1931
63. HUXLEY, T. H., *Science and Education*, London, 1925
64. JACKSON, B. D., *Linnaeus*, London, 1923
65. JOSEPHSON, M., *Edison: a Biography*, London, 1961
66. KENT, A. (ed.), *An Eighteenth Century Lectureship in Chemistry*, Glasgow, 1950
67. KNOWLES, L. C. A., *The Industrial and Commercial Revolution in Great Britain During the Nineteenth Century*, London, 1941
68. LAGRANGE, J. L., *Mécanique Analytique*, Paris, 1788
69. LARGE, E. C., *The Advance of the Fungi*, London, 1940
70. LAVOISIER, A. L., *Oeuvres*, 6 vols., Paris, 1864–93
71. LYELL, C., *Principles of Geology*, 3 vols., 1830–33
72. MACH, E., *The Science of Mechanics*, 5th ed., London, 1942
73. MCKENDRICK, J. G., *Hermann Ludwig Ferdinand von Helmholtz*, London, 1899
74. MCKIE, D., *Antoine Lavoisier*, London, 1952
75. MCKIE, D., and HEATHCOTE, H. N. DE V., *The Discovery of the Specific and Latent Heats*, London, 1935
76. MALTHUS, T. R., *An Essay on the Principle of Population*, 6th ed., London, 1826
77. MARTIN, K., *French Liberal Thought in the Eighteenth Century*, London, 1954
78. MARTIN, T., *Faraday's Discovery of Electro-Magnetic Induction*, London, 1949
79. MARX, K., *Capital*, vol. I, London, 1946; vol. II, Chicago, 1885; vol. III, Chicago, 1909
80. MARX, K., *Selected Works*, vol. I, London, 1942

81. MARX, K., *Selected Works*, vol. II, London, 1942
82. MASON, S. F., *A History of the Sciences*, London, 1953
83. MAXWELL, C. (ed.), *The Scientific Papers of the Hon. H. Cavendish*, 2 vols., Cambridge, 1921
84. MEIKLEHAM, R. S., *Descriptive History of the Steam Engine*, London, 1824
85. MENSHUTKIN, B. N., *Russia's Lomonosov*, Oxford, 1952
86. MEYER, R. W., *Leibnitz and the Seventeenth-century Revolution*, Cambridge, 1952
87. MORIN, J. B., *Astrologia Gallica*, The Hague, 1661
88. MOURET, G., *Sadi Carnot et la Science de l'Énergie*, Paris, 1892
89. MURRAY, R. H., *Science and Scientists in the Nineteenth Century*, London, 1925
90. NASMYTH, J., *Autobiography*, London, 1883
91. NICOLLE, J., *Louis Pasteur: a Master of Scientific Enquiry*, London, 1961
92. POLHAMMER, C., *Patriotic Testament*, London, 1761
93. PRICE, D. J. DE S., *Little Science, Big Science*, New York, 1963
94. PRIESTLEY, J., *Experiments and Observations on Different Kinds of Air*, 3 vols., London, 1775–7
95. PRIESTLEY, J., *The History and Present State of Discoveries Relating to Vision, Light and Colours*, London, 1772
96. PRIESTLEY, J., *The History and Present State of Electricity*, London, 1775
97. PRIESTLEY, J., *Memoirs*, London, 1806
98. RAISTRICK, A., *Dynasty of Iron Founders: The Darbys and Coalbrookdale*, London, 1953
99. RAMSAY, Sir W., *The Life and Letters of Joseph Black*, London, 1918
100. REYNOLDS, O., *Memoir of James Prescott Joule*, Manchester, 1892
101. RICARDO, D., *Principles of Political Economy and Taxation*, London, 1924
102. RUNES, D. D. (ed.), *The Diary and Sundry Observations of Thomas Alva Edison*, New York, 1948
103. SALUCES, M. DE LUR, *Lomonossof*, Paris, 1933
104. SAVERY, T., 'The Miners' Friend' (1702), *Supplement to the Series of Letters Patent, etc. (1617–1852)*, vol. I, London, 1858
105. SCHOFIELD, R. E., *The Lunar Society of Birmingham*, Oxford, 1963
106. SCHUBERT, H. R., *History of the British Iron and Steel Industry from c. 450 B.C. – A.D. 1775*, London, 1958
107. SCOTT, Sir S. H., *The Exemplary Mr Day*, London, 1935
108. SEGUIN, M., *De L'Influence des Chemins de Fers, etc.*, Paris, 1839
109. SHERRINGTON, Sir C. S., *The Integrative Action of the Nervous System*, Cambridge, 1947
110. SILK, L. S., *The Research Revolution*, New York, 1960
111. SIMON, J., *English Sanitary Institutions*, London, 1890
112. SMILES, A., *Samuel Smiles and His Surroundings*, London, 1956
113. SMILES, S., *Industrial Biography*, London, 1908
114. SMILES, S., *Lives of the Engineers*, 5 vols., London, 1904
115. SMILES, S., *Self Help*, London, 1950
116. SMITH, A., *An Inquiry into the Nature and Causes of the Wealth of Nations*, 2nd ed., Edinburgh, 1846
117. SNOW, Sir C., *The Two Cultures and the Scientific Revolution*, Cambridge, 1959
118. SOCIÉTÉ FRANÇAISE DE MINÉRALOGIE, *René-Just Haüy*, Paris, 1945
119. STRACHEY, J., *The End of Empire*, London, 1959
120. SWAN, Sir K. R., *Sir Joseph Swan*, London, 1946

121. TATON, R., 'The French Revolution and the Progress of Science', *Centaurus*, vol. 3, 1953

122. THOMAS, R. H., *Liberalism, Nationalism and the German Intellectuals 1822–47*, Cambridge, 1952

123. THOMPSON, S. P., *The Life of William Thomson*, 2 vols., London, 1910

124. THOMPSON, L. G., *Sidney Gilchrist Thomas*, London, 1940

125. THOMSON, Sir W., *Mathematical and Physical Papers*, vol. 5, Cambridge, 1911

126. THORPE, Sir T. E., *Joseph Priestley*, London, 1906

127. THURSTON, R. H., *A History of the Growth of the Steam Engine*, New York, 1939

128. TILLYARD, A. L., *A History of University Reform*, Cambridge, 1913

129. TOYNBEE, A., *Lectures on the Industrial Revolution in England*, London, 1844

130. TRENEER, A., *The Mercurial Chemist: a Life of Sir Humphry Davy*, London, 1963

131. TUGE, H., *Historical Development of Science and Technology in Japan*, Tokyo, 1961

132. UNITED NATIONS, *Science and Technology for Development*, 8 vols., New York, 1963–4

133. URE, A., *The Philosophy of Manufactures*, London, 1835

134. VALLERY-RADOT, R., *The Life of Pasteur*, London, 1920

135. VAVILOV, S. I., *Soviet Science: Thirty Years*, Moscow, 1948

136. WILKINSON, C. S., *Wake of the Bounty*, London, 1953

137. WOODWARD, E. L., *The Age of Reform: 1815–1870*, Oxford, 1938

138. WOOLF, L. S., *After the Deluge*, 3 vols., London, 1931–53

Note on the Illustrations

The choice of illustrations for Professor Bernal's *Science in History* has been based on the simple principle of providing additional illumination of the text. Since the author has taken so wide a canvas on which to display his analysis, the range of illustrations has accordingly been made as broad as possible. However, science has not always been illustrated at every stage in its history and from some periods, of the few illustrations which may have existed, little or no evidence has survived to the present day; in consequence certain problems had to be solved if gaps were to be avoided. For example, virtually no original material remains of Greek science, and the scientific texts that we have are copies or translations made in later centuries. In such cases, later sources have been used if, as often happens, they make the point; Greek ideas continued for so long in western Europe that it is often still valid to use material from printed books.

In this book, where both science and the interplay of social conditions are discussed, the pictures could not always be chosen as direct illustrations of the text, but in every case it is hoped that the full captions will enable the reader to see why a picture has been chosen and appreciate its relevance, whether as allusion or analogy, by comparison or even as a comment. No attempt has been made to illustrate Professor Bernal's introductions to the various sections of his book, since this would have caused too great a mixture of varying subjects and historical periods. By confining illustration to the main body of the text, some degree of chronological order has been possible.

The choice of each picture has depended on a number of factors: its relevance to the text, the quality of the illustration itself, its power to provide additional visual or factual information and, of course, its aesthetic appeal. Here and there diagrams have been used, but in every case they are of historical significance. In volume 1, except for the need to cast the net wide for material about Greek science, the illustrations are comparatively straightforward. Volume 2 has almost illustrated itself. Volume 3, dealing primarily with modern scientific research, is again straightforward, but volume 4 has presented some problems, in that its theme – the social sciences in history – is so wide, and that some of the concepts cannot be illustrated directly. The solution adopted has been to try, in one way or another, to complement the spirit of the text. Sources of illustrations have been given wherever possible, in a separate acknowledgements section on p. 693.

My thanks are due to Mr Francis Aprahamian for his helpful advice, and especially to my wife, whose assistance and extensive library of illustrations has proved invaluable.

COLIN A. RONAN
Cowlinge, Suffolk
June 1968

Acknowledgements for Illustrations

For permission to use illustrations in this volume, acknowledgement is made to the following: Biblioteca Ambrosiana, Milan for number 109; the Bodleian Library, Oxford, 132; the Trustees of the British Museum, 116; John Freeman & Co. Ltd, 202; E. P. Goldschmidt & Co. Ltd, 123; the Lowell Observatory, 145; the Mansell Collection, 164; the National Maritime Museum, 146a, 146b; Thomas Nelson & Sons, 200; Ronan Picture Library, 101, 102, 103, 104, 105, 106, 110, 111, 112, 113, 115, 117, 121, 124, 126, 127, 135, 136, 137, 140, 143, 144, 147, 148, 149, 150, 151, 152, 153, 154, 155, 156, 157, 158, 160, 161a, 161b, 162, 163, 165, 166, 168, 169, 170, 171, 172, 173, 175, 176, 177, 178, 179, 180, 181, 182, 184, 186, 187, 188, 189, 190, 191, 193, 194, 195, 196, 197, 198, 199, 201, 203, 204, 205, 206; Ronan Picture Library and E. P. Goldschmidt & Co. Ltd, 107, 108, 122, 128, 129, 130, 134, 138, 141, 183; Ronan Picture Library and the A. R. Michaelis Collection, 159; Ronan Picture Library and the Royal Astronomical Society, 114, 118, 119, 120, 125, 133, 139, 142; the Managers of the Royal Institution, 185; the Royal Society, 131; the Science Museum, 174, 192; the Trustees of the Tate Gallery, 167.

The publishers would like to thank Mrs Sheila Waters for drawing map 4.

Name Index

Vol. 1: 1–364   Vol. 2: 365–694   Vol. 3: 695–1008   Vol. 4: 1009–1330

Subject Index

Bold figures indicate main reference

Vol. 1: 1–364   Vol. 2: 365–694   Vol. 3: 695–1008   Vol. 4: 1009–1330

Abacus, 120, 276, 333
Abbasids, 270, 272, 276, 349
Academia Sinica, 710
Academic freedom, 844 f., 1254, 1260 f.
Académie Royale des Sciences, 448, 450, **451**, 455, 497, 498, 517, 534, 537
Academies: in ancient world, 196 f., 207, 212 f.; eighteenth- and nineteenth-century, 514 f., 529, 549 f., 675; Renaissance, 451; seventeenth-century **450** ff.; Soviet, 709, 1190, **1265** ff. *See also* Royal Society, Académie Royale, etc.
Academy, The, **196** ff., 207, 450
Accademia del Cimento, 451 f., 455
Accademia dei Lincei, 451
Accelerator, particle, 714, 752 ff., 764 f., 843, 849 ff., 994
Advertising, 1142 f.
Aerodynamics, 811
Aeroplane, 712 f., **809–13**, 1257
Africa, 89, 140, 145, 153, 267, 269, 276, 281 f., 402, 707, 710, 723, 948, 1027, 1120, 1174, 1200 f., 1202 f., 1251 f.
Age of Reason, **531** f., 544, 1056
Agricultural Revolution, **524** f.
Agriculture: medieval, 287 f., 291 f., 312, 679; modern, 880 f., 970 ff.; nineteenth-century, 565 ff., 654 f.; origin of, 16, **91–6**, 99 ff., 342 f., 716; Roman, 224; seventeenth- and eighteenth-century, 511, **524** f.
Air-pump, 470 f.
Aix-en-Provence, 451
Akkad, 136
Albigenses, 294
Alchemy: and chemistry, 398 f., 439, 472 ff., **617** ff.; Chinese, 279 f., 1226; Islamic, 279 f.; medieval, 301, 306,

310; modern, 734 f.; origins of, 128, **222**; Renaissance, 324, 398
Alcohol, 279 f., **323** ff., 918
Alexandria, 161, 169, 209, 212 ff., 223, 227, 251, 257, 258, 270; Great Library, 231; Museum, 169, **212** f., 217, 223, 241, 272, 295, 450
Algebra, 122, 276, 332, 443
Algeria, 1120, 1130, 1180, 1199
Alpha particle, 735, 738, 752, 754 ff., 763
Alphabet, 156 f.
Alum, monopoly, 399
America: Philosophical Society, 526; War of Independence, 529, 805, 1053, 1057, 1317. *See also* United States, Latin America
American Indians, 70, 125, 535, 1032, 1241
Amplifiers, 776
Anarchism, 1108
Anatolia, 162, 342
Anatomy, 125, 188, 223, **392** f., **437**, 469, 645 f.
Animal behaviour, 71, **944** ff.
Antarctica, 799, 831
Anthrax, 565, 651
Anthropology, 1019, **1081** f., 1157, 1181 f., 1208
Antibiotics, 877, 924 ff., 929, 980
Anti-Communism, 16, 1122, 1124, 1128, 1130, 1132, 1137, 1157, 1264
Anti-ferromagnetism, 793, 854
Antioch, 251, 257, 270
Anti-particles, 764 ff., 767
Anti-Semitism, 1129, 1130
Apocalypse, 233, 402
Arabic numerals, 122, 264, 332, 1226
Arabs, 197, 209, 217, 253, 258, 266,

École de Médecine, 536
École Normale Supérieure, 536, 553, 648
École Polytechnique, 536, 553
Ecology, 960–72 *passim*
Economics, 1036, 1044, **1047–53** *passim*, **1059** ff.; Keynesian, 1090, 1133ff.; marginal theory, 1087 ff.; Marxist, **1067–78**, 1100 ff., 1166 f.; Soviet, 1183 ff.
Education, 17, 74; in antiquity, 173, 191 ff., **196** f., 198, 207; in Bronze Age, 130 f.; in China, 1198 f.; eighteenth-century, 529 f., 536; medieval, **295** ff.; nineteenth-century, 552 ff.; primitive, 74; science of **1148–52**;, in the Soviet Union, 1153, **1187** ff.; technical, 552 f., 1153; higher, 1150 ff., 1264 f., 1275
Egypt: anti-Hellenic movement in, 263; early civilization, 62, 100 f., **104** ff., 109 112, 114–27 *passim*, 139, 157, 161 u., 212, 279, 385; early science and technique, 112, 119-48 *passim*, 163 f., 177, 218, 248, 347; modern, 967, 969, 1120, 1201
Electrical discharge, **601** ff., 617 f., 730
Electrical industry, **563** f., 705, 729, 753, 805, 1254
Electricity: eighteenth-century, 516, **600–606**; nineteenth-century, **607–17**; lags in application of, 612 f.
Electro-encephalograph, 885, 942
Electrolysis, 627
Electromagnetic theory of light, **567** f., 574, 744, 1228
Electromagnetism, 607
Electron, 280, 473, 620, 627, 730, **732** f., 737, 739 f., 748, 751, 755, 768, 775, 780, **785** f., 793, 795; microscope, **785** ff., 873, 885 f., 931, 982
Electronic computers, 10, **783** f., 817 f., 825, 856, 915, 982, 1145 f., 1281; theory of metals, 795 f.
Electronics, **773–87**
Elements: Boyle's definition of, 473; establishment of chemical, **623** f.; Greek, 78, **174** f., 188 f., 199 f.; Islamic, 280; periodic table of, 568; Renaissance, 398, 439
Élite, intellectual, 11, 130 f., 166, 212, 1033 f., 1150, 1-188, 1196, 1198, 1275, 1278, 1285 f., 1287 f.
Embryology, 647, 930–36 *passim*
Empire, **138** f., 190, 208 ff., 224 ff., 251 ff., 269 ff. *See also* Imperialism
Enclosure Acts, 511

Encyclopédie, 306, 526 f.
Endocrinology, 899, **939** f.
Engineering, 41 f., 218, 547; nineteenth-century, **590–98**; twentieth-century, **804–23**
Engineers, 41 f., **137** f., 218, **395** f., 521, **547**, **591** f.
England: eighteenth-century, 509, 511, 512 ff.; Industrial Revolution, 47, 413, 414; medieval, 286, 291, 1220; nineteenth-century, 552 f., 626, 631; Renaissance, 403, 418; seventeenth-century, 441, 454, 490, 495, 497, 504, 1243; twentieth-century, 17, 1153
Enlightenment, The, 489, 540, 543, 553, 1247
Enzymes, 649, 796, **887** ff., **892** ff., 903, 907 f.
Epicureanism, 212, 227, 1037
Epidemic diseases, 565, **651** f., **879** f., 912, 977 ff.
Eskimos, 69, 83 f.
Essenes, 158, 255
Ether, 199, 612, 744
Etruscans, 210
Eugenics, **1085** f., 1140
Evolution: **556** ff., 575, **640–45**, 662 f., 986; twentieth-century, **947–60** *passim*; and social science, 1084 f.
Existentialism, 227, 1131 f.
Exodus, 1032
Expanding universe, 770 f.

Fabian Society, The, 1097 ff.
Fabianism, **1097** ff., 1111 f., 1119, 1124, 1140, 1167
FAO, 897, 973, 974
Fascism, 195, 1119, 1123, 1129 f., 1136, 1157, 1211. *See also* Nazism
Fermentation, 70, **399**, **647** ff., 877, 887
Fertilizers, 566, **655**, 657, 823 f., 881, 970 f., 974, 1304
Feudal: system, **285–335** *passim*, 350, 1242 f.; serfs, 288; village, 287 f.
Final causes, **201** f., 207, 213, 442, 558
Fire, **69** f., 81, 88, 91, 133, 322, 1229. *See also* Combustion
Fireworks, 237, 814
First International, The, 1105, 1108 f.
Flanders, 287, 289, 329, 389, 474, 1242
Flatworms, 945 f.
Flood, Noah's, 636, 640
Florence, 5, 329, 379, 386, 396, 400
Food-gathering, **74** f., 84
Food industry, 655 f., 878 f.
Food supply, 869 f., 885, **973** ff., 988, 1301 ff, 1305 f.